Healthier Workers

Healthier Workers

*Health Promotion and
Employee Assistance Programs*

Martin Shain
Helen Suurvali
Marie Boutilier
Addiction Research Foundation of Ontario

Lexington Books
D.C. Heath and Company/Lexington, Massachusetts/Toronto

Library of Congress Cataloging-in-Publication Data
Shain, Martin.
  Healthier workers.

  Bibliography: p.
  Includes index.
  1. Employee assistance programs. 2. Preventative health services. 3. Alcoholism and industry. 4. Health promotion. I. Suurvali, Helen.
II. Boutilier, Marie. III. Title.
HF5549.5.E42S52   1986   658.3′82   85-45038
ISBN 0-669-09908-2 (alk. paper)

Published simultaneously in Canada
Printed in the United States of America
International Standard Book Number: 0-669-09908-2
Library of Congress Catalog Card Number: 85-45038

The paper used in this publication meets the minimum requirements of American National Standard for Information Sciences—Permanence of Paper for Printed Library Materials, ANSI Z39.48-1984.
(∞)™

The last numbers on the right below indicate the number and date of printing.

10 9 8 7 6 5 4 3 2 1

95 94 93 92 91 90 89 88 87 86

Contents

Figures and Tables

Figures

Tables

Acknowledgments

The stress management course that is evaluated in the second part of chapter 6 was designed and implemented by Fern Gue in collaboration with James Simon. The authors acknowledge Fern's contribution to this chapter with thanks.

Robyn Lim was instrumental in deciding the shape of chapter 4 by conducting research that ruled out a number of false leads. We appreciate her behind-the-scenes diligence.

Richard Miller, currently manager of the Health Management Program at Xerox, and Keith McClellan, director of the Tri-County EAP in Akron, Ohio, both deserve our thanks for their encouragement of the view that employee assistance and health promotion professionals have something to offer one another.

We are most grateful to Howard Grimes, director of the Center for Work Performance Problems and to Dick Groepper, director of Employee Assistance Services, Crawford and Company, both of Atlanta, Georgia, for providing an opportunity to test the central thesis of this book in the forum of the Southeastern EAP Institute.

We owe a debt of gratitude to Lecia Hanycz, our secretary, for her cheerfulness and equanimity under the pressure of producing the manuscript for this book on time.

Introduction

This book is an attempt to place two vitally important health interventions, employee assistance programs (EAPs) and health promotion programs (HPPs), into a single philosophical, theoretical, and empirical framework that will allow practitioners and researchers in both areas to see potential for a greater synergy between them. This undertaking arose from the observation that the professionals who ply their trades in each of these fields are often very different sorts of people with widely varying backgrounds and disciplinary affiliations. So great is this diversity that the opportunity for collaborative work and for comprehensive planning is often missed because it seems to be assumed that EAP and HPP are separate domains with little to unite them. We argue in the following chapters that the unifying factor is to be found in the health needs of workers. These needs fall along a spectrum from health maintenance and promotion to health restoration and rehabilitation. While it is often assumed that HPPs and EAPs operate at the polar ends of this spectrum, we can see potential for outreach by both forms of intervention so that the needs of the whole work force are accommodated. We argue in a case that is summarized in chapter 7 that both employees and employers stand to benefit from such outreach.

Nonetheless, these program extensions must be developed according to a set of principles that recognizes the influence of environmental factors in the genesis and amelioration of physical and mental health problems. In addition, components must be built into such programs that foster in participants or clients a sense that they have some, albeit limited control over these factors, be they in the workplace, in the family, community or larger society. The accomplishment of this task, diffuse and sprawling though it is, will occur only if HPP and EAP practitioners accept the need for it and for the extended roles that it requires them to play. Thus, part of the book is an attempt to explore the impact of social, economic, and organizational constraints upon individual wellness and health-related choices in such a way that implications for feasible extensions of EAP/HPP roles can be drawn out.

We hope that the somewhat critical approach that we have taken to EAPs and HPPs will not be perceived as gratuitous negativity. We believe on the basis of evidence that we have seen and obtained through our own research reported here that components can be added to EAPs and HPPs that will allow them to better serve the needs of employees as well as to continue serving the needs of employers. This position is advanced from a perspective in which the work of EAP and HPP professionals is highly valued and respected by the authors. We presume to offer suggestions for future developments of their work only in this spirit of respect and in the knowledge that EAP and HPP professionals are by and large a group who welcome challenge and the possibility of new horizons.

1
Employee Assistance and Health Promotion Programs: Implications of the Past for Practitioners Today

with Christine Shain

The purpose of this chapter is to demonstrate that the history of employee assistance and health promotion in the workplace has much to teach us about the real nature of the problems faced by practitioners associated with these fields today and about how some of these problems might be confronted. This exercise will serve to presage one of the principal themes of this book, which is an examination of the roles of employee assistance and health promotion professionals in the workplace particularly with regard to whom they really serve: Is it the worker, the employer, or both? This is not a new issue in either employee assistance or health promotion. With regard to the former, Briar and Vinet (1985) have recently argued that although the dictates of sponsorship and auspice make it difficult, the EAP professional with social work affiliations should "be motivated by values that prompt advocacy and promote equity" (p. 347). In so saying, Briar and Vinet reflect Ozawa's earlier exhortation to industrial social workers that they be involved in attempting to identify and solve structural problems—that is, those problems pertaining to the organization of work that affect the mental and physical health of workers (Ozawa 1980). With regard to health promotion, Navarro (1976), Labonté and Penfold (1981), and Crawford (1977; 1980) among others have pointed out that it is ethically questionable to run programs that implicitly or explicitly hold individuals completely responsible for their deteriorating health and for its restoration in situations where it is clear that the employer could do much to create a more salutary environment in terms of both mental and physical well-being. These commentators argue that "victim blaming" by default places the health programmer in a position where he or she is clearly the agent of the employer, not the employee.

An examination of the historical roots of employee assistance and health promotion will hopefully lead us away from the sanctimony that can on occasion attend the criticisms leveled at practitioners in these areas. A certain self-righteousness sometimes afflicts those who feel that such professionals act more as employer lackeys than worker advocates. This attitude may reflect a lack of awareness concerning the difficult birth of health-related programs or interventions in the workplace and from the fact that modern programs, for all their aspirations, are tied to long histories that have shaped their orientation toward worker needs. Putting aside philosophical conflicts, there is a need to come to terms with the past by recognizing its power and by systematically loosening its grip so that the emerging professional standards of employee assistance and health promotion practitioners can jockey for position with the legitimate demands of employers for stable, productive, and profitable work forces.

Views from the Past

Despite current concerns about the flirtation between modern practitioners of employee assistance programs (EAPs) and health promotion programs (HPPs) as presented by Roman (1981b; 1984), the two forms of intervention have a common origin if one goes back far enough. This is not to suggest that the current concerns are without foundation since today the people who run EAPs and those who run HPPs are usually quite different sorts of people with varying philosophies, theories, and practices (these varying perspectives are further reviewed in subsequent chapters, particularly 3 and 7). At one time, however, the very concept of "program" was quite alien. What we know as EAPs and HPPs today would, when they existed at all, have been subsumed under the general auspices of working conditions created by autonomous and autocratic employers. Thus, the history of EAPs and HPPs is also the history of "employer–employee" or more legalistically of "master–servant" relationships.

The evolution of this relationship in the last millennium has been at times quiet and at others stormy. An arbitrary but convenient place to begin an inquiry is the early Middle Ages where we will find that the dominant form of employment was within the framework of the feudal system (Poole 1955; North and Thomas 1971; Fenoaltea 1975). The key characteristic of this system, for modern observers of EAP and HPP was that protection, privileges, and sometimes remuneration were afforded by feudal landlords to subordinates of various degrees in return for their fealty, which they provided in the form of services, goods, tithes, and taxes. The lower the social order the less choice there was about these arrangements.

As the coherence of the feudal system began to dissolve, a residual system was left behind in the form of paternalism. The dictionary defines this system as one "under which an authority undertakes to supply needs or regulate conduct of those under its control in matters affecting them as individuals as well as in their relations to authority and to each other." In this system, the economic relationship between employer and employee existed in the form of a contract regarding the exchange of money for services. In theory, breaches of such contracts were actionable by common law. However, much was left to the goodwill of the employer with regard to the provision of decent wages, working conditions, and benefits. Needless to say, the range of such agreements was enormous. Examples can be found of situations that were essentially servile and of others in which the workers were provided voluntary pension and retirement benefits, including a place to live. The latter type of situation was never the prevailing norm, although by the mid-nineteenth century numerous examples of benevolent paternalism could be found.

The trend toward employer benevolence, which can be detected prior to or in addition to mandating legislation or union advocacy, had more than one root itself. In his examination of the growth of industrialism in the United States between 1885 and 1914, Hays (1957) provides a context for understanding the climate in which industrial welfarism or American welfare capitalism developed. He notes that the nineteenth century was characterized by the countervailing forces of liberal, utopian socialist, philanthropic movements on the one hand (particularly manifest in the form of Christian socialism) and variations upon the theme of social Darwinism on the other. The jostling between these forces was nowhere more visible than in the workplace, yet it is often assumed that the paternalism that typified industry at the time was monochromatic in so far as employers were seen to be relentlessly calculative and ingeniously exploitive in their use of welfare policies. It is often argued, for example, that one of the principal motives for the development of American welfare capitalism was the desire to prevent unionization (Brandes 1976). Clearly this was the case in some, perhaps many, instances. However, examples can be found of employer policies that fall on all points of a spectrum ranging from despotic paternalism on the one hand to an almost participatory, philanthropic paternalism on the other. Case studies of various employer policies are sketched in the following section. These descriptions serve to illustrate how virtually the whole range of philosophies about health and well-being that existed in the early part of this century and before can be detected in various guises in the employee assistance and health promotion programs of today.

Examples of Industrial Welfarism

The Use of Welfare Secretaries

According to Popple (1981), 33 percent of the 431 largest companies in the United States had at least one full-time welfare secretary by 1919, based on

Bureau of Labor Statistics data. Another 36 percent contracted with outside agencies for social work services. While the welfare secretary was really only a small part of the industrial welfare approach, a disproportionate amount of attention has been paid to this office (Popple 1981; Bargal and Shamir 1982; Thomlison 1983). Brandes (1976) has supplied, in the form of his treatise on American welfare capitalism, an excellent framework for the study of the services provided by these forerunners of modern industrial social workers. Popple suggests that the whole purpose of industrial welfarism in general and of the welfare secretary in particular was to bring workers in line with management values and ideals, thus eliminating the desire to unionize. Indeed, union leaders of the time tended to feel that welfarism was an attempt to buy off workers and to distract them from perceiving the need for more basic changes in the relations between capital and labor (Popple 1981).

It seems clear, however, that the welfare secretary was employed for a variety of reasons. In the earliest least formal context she operated as a "house mother" for employees—supervising their living arrangements, listening to their problems and offering advice, and acting in loco parentis with respect to moral standards. Between the years 1818 to 1850 the cottonmills of Lowell, Massachusetts were able to attract a work force of middle-class young American women because of the familial environment maintained by the company agents, promising security to the girls and particularly to the parents who allowed them to go (Josephson 1949; 1967). In other situations her intervention was more specifically oriented and she operated somewhat like a public health nurse today. In a more expanded context, she (or sometimes he) served the now traditional social work roles of service broker, intermediary, supporter, advocate, and therapist but sometimes found herself too in the position of recommending disciplinary action and even dismissal. In the latter sense, and in so far as she managed benefits such as pensions, insurance, clubs, lunches, clinics, and so on, she was the precursor of the modern personnel manager. The variety of roles that welfare secretaries played served to place them frequently in conflict-of-interest situations as far as workers were concerned. This fact probably led to the decline in the use of such secretaries, although it is hard to distinguish their fate from that of industrial welfarism in general, which came into greater disfavor with the approach of the 1930s. Thus, the early welfare secretary's variety of roles served to illustrate the question that is still being asked today, namely: Who is the client—the worker or the employer? Popple (1981) suggests, with Briar and Vinet (1985) that the modern social worker tends to favor the worker's interests by virtue of the standards and codes of ethics that are coming to characterize the profession. Social work, however, is only one of the professions within the varied context of EAPs and does not necessarily represent the posture of the field as a whole.

Industrial Welfarism in General

The welfare secretary was but one expression of a whole movement within industry that was guided in one form or another by a paternalistic philosophy. At its most comprehensive, welfarism in industry embraced most areas of critical importance to employee well-being: housing, medical care, overall fitness, nutrition, recreation, savings, pensions, education, and general working conditions. Sometimes these benefits were provided as expressions of an integrated world view that saw individual, organizational, and social health as inextricably linked, while in other cases they were calculatingly provided as lubrication to workers as machines that needed optimal conditions to perform at their best. The vision that guided some of the experiments in industrial welfarism rivals such apparently modern approaches as social technical innovations or quality of working life programs. The principal difference lies in the degree to which workers are consulted about the process of change. However, even today the participation of workers in such grand schemes is often more imaginary than real (see chapter 8 for elaboration).

Welfarism was in many respects *preventive* in its orientation; it was a structural approach to the prevention of problems that might affect job performance. Today, it is commonplace for EAPs to deal with employee problems related to housing or accommodation and with the stress due to domestic and financial worries or daycare for children. HPPs deal with problems of nutrition, fitness, stress, and the identification and remediation of personal habits that adversely affect health. Both types of program deal remedially or preventively with alcohol and drug problems. Industrial welfarism, in its fully developed form, was based on the view (whether calculating or benevolent in origin) that these problems, which we now tend to deal with programmatically, could be prevented by properly ordering the lives of employees and addressing their most basic needs. This was usually predicated on the assumption that employers knew what was best for their employees. Nonetheless, it is worth observing that in the comprehensiveness of its embrace, industrial welfarism incorporated many elements of what are now quite distinct programs in employee assistance and health promotion.

The Ford Approach

In relation to the spectrum of paternalism referred to earlier, Henry Ford provides us with an excellent example of benevolent despotism. Ford has been described variously as tyrannical, innovative, courageous, erratic, eccentric, brilliant, and totally nonintrospective (Nye 1979; Gelderman 1981). His world view appears to have been highly idiosyncratic, possessing little internal consistency or integration. It was also notably secular. Ford believed that the machine was the new Messiah (Gelderman 1981, 129), and workers were organic machines for which he provided optimal maintenance conditions. Character, crime, and disease were seen as products of variations in diet; consequently, at least for his executives, he prescribed dietary regimens that were heavy on soybean products, salads, and "roadside greens" but omitted liquor, coffee, doughnuts, and cookies.

Executives were expected to maintain a high level of fitness for which Ford provided a comprehensive dancing program and an on-site ballroom. He employed full-time instructors for executives, who were expected to attend Friday night dance parties with their wives. At these dance parties, which were a fixture of Ford executive life from 1924 to 1943, no sitting down was allowed (save for one old friend who had had two strokes and was confined to a wheelchair), and if there was an odd man out, he was constrained to dance by himself. Dancing was the preferred mode of exercise: Ford considered squash and handball pretentious. At the end of lunch meetings it was not uncommon for Ford to be taken with a desire to "get jigged up a bit," thus causing the assembled group to engage in a little old-time dancing for an hour or so (Gelderman 1981, 281). At other times, Ford would suddenly challenge his executives to "high kicking" contests as tests of their physical condition.

Ford had less success in subjecting the rank and file to his fitness and health directives, although in the early years workers had to allow themselves to be governed in all areas of their lives if they expected to be promoted or in some cases even to keep their jobs. In 1914, the short-lived Sociological Department was formed to promote a "scientific" model of home life. It set up standards of domestic hygiene, interior decoration, and healthy family functioning according to which eligibility for certain benefits was determined. Renting rooms to boarders was frowned upon, offenders being disqualified from the higher pay scales.

Ford strongly espoused an industrial model that would allow workers time to farm as well as hold down factory jobs. Workers could save money by growing their own food, gain exercise, and keep in touch with the soil. His early attempts to encourage his workers to pursue this philosophy in an urban setting met with little success. Ford persisted, however, in his belief that the success of the American industrial state depended on a melding of rural virtues and industrial efficiency: "With one foot in the land and one foot in industry, America is safe" (Gelderman 1981, 290). After some experimentation, Ford established the Ford Village Industries. Their rural locations, in modernized former mills, and smaller work forces allowed time sharing and leaves of absence so that workers could farm.

On the shop floor, the service department saw to it that efficiency was the byword. Silence was enforced, rest and meal times were extremely short, and a system of constant surveillance and spies led to frequent dismissals for real or alleged offenses. By 1932 such conditions led to a strike and the infamous Dearborn massacre that was perpetrated to crush it. This gave way to an even worse period. The years 1937–1940 were known as the "Ford Terror," an era during which the service department exercised tyrranical control. For all this, Ford was a seductive employer, at least in the early years. His "Five Dollar Day" introduced in 1914 effectively doubled workers' wages overnight, much to the disgust of the business community and unions alike who envisioned a move toward profit sharing.

It appears that Ford believed his social philosophy, once implemented, provided working conditions that were in the best interests of workers as well as profitable for him. Nonetheless, it is clear that no financial benefit could compensate for the indignity of being treated as a machine in the service of an industrial mogul. Extreme though this case may appear, we refer in later sections of this book to modern expressions of Ford's world view, which, though well disguised, are based upon the premise that workers are production units that need to be well maintained in order to function well.

The Case of Amoskeag

The Amoskeag Textile Mills operated between 1838 and 1936 in and around Manchester, Massachusetts. It was a consolidation of thirty smaller textile mills with 17,000 employees at its peak. The firm was run by agents of the Amoskeag Board whose members lived mostly in Boston. This organization developed a "corporate welfare and efficiency" program that had as its primary objectives the attraction of immigrant workers (who would then be Americanized), the instillation of loyalty, the curbing of labor unrest, and the prevention of unionization (Hareven and Langenbach 1978). As such, Amoskeag was among the more calculative companies in its use of welfarism. This was no externalized, projected personal ideology but rather the deliberate marketing of what was seen to be a serviceable idea. The company was relatively self-sufficient in that it had its own source of power and developed its own carefully planned community of private houses, lodging, businesses, churches, clubs, parks, and gardens. Most of the land was put up for sale to employees with more than five years seniority. Mortgages were available at extremely favorable rates. The quality of the accommodation was, by all accounts, of a superior type and the environment beautifully maintained until the 1920s when competition from the South began to erode the system.

Originally, the majority of the work force was made up of young, unmarried women and girls for whom Amoskeag acted in loco parentis. Meals were provided at low cost in boarding houses where curfews were imposed. Church attendance was made compulsory, alcohol consumption was prohibited, and smoking forbidden as a safety measure. Emergency health care was always available on site, and two nurses were on duty mornings for other health concerns. A physician was on call in the afternoon and was able to make house calls along with five visiting nurses. A free dental clinic was established around 1910. The Amoskeag Textile Club consolidated and expanded from about the same date a wide variety of recreational, athletic, and social activities that had hitherto been carried on independently. The Amoskeag Domestic Science School operated from at least 1880, offering cooking and sewing classes to employees, while the Textile Schools were established in the early years of the twentieth century to provide technical training in job-related areas for motivated individuals.

Amoskeag appears to be one of the earliest documented examples of "canned welfarism"—that is, welfare applied for the sole purpose of increasing productivity and reducing labor problems. In this kind of environmental-control model of welfarism, workers were expected to respond favorably to the conditions created for them. Dismissal came easily because it was assumed that no one in their right mind would bite the generous hand that fed them. If the social experiment failed, the worker, not the employer, was to blame. The common feature between the Ford/Amoskeag approach and some modern HPPs is the lack of consultation with employees about their needs and how they should best be accommodated in the framework of an industrial situation (see chapters 2, 7, and 8 for elaboration of this argument).

Goodyear

The philosophy of Frank and Charles Seiberling, who in 1898 founded Goodyear in Akron, Ohio, appears to have been that if you hired healthy people and kept them healthy, you would have a healthy organization. Team spirit was highly prized and hiring practices favored the recruitment of athletes, particularly for leadership positions, and of Boy Scouts. Goodyear, under the leadership of P. W. Litchfield, an early superintendent and later chairman of the board, was extremely active in promoting the Boy Scout movement. Scouts working for Goodyear were provided with special incentives, while staff from Litchfield down to foreman were encouraged to become active in Scout leadership (Allen 1949). In 1909, prior to legislation requiring it, Goodyear formed the Relief Association to administer the collection of funds and disbursement of benefits for sickness and injury. The company had the second earliest factory hospital in the United States (the first being International Harvester's) with six beds and employing a full-time doctor with day and night nursing. By 1927, care in local hospitals was available through contracts undertaken with the Relief Association. Within the first twenty years of operation, Goodyear provided paid vacations, an eight-hour day, and a pension plan.

In 1912, the firm built a planned community on one hundred acres called Goodyear Heights just north of Akron. It was self-sustaining with regard to water, sewage, and power. Housing of superior quality was available for purchase by employees at a reasonable cost. It was built against the advice of superintendent Litchfield. He argued that housing was no business of the company and that in other company towns it had inspired employee complaints because of the belief that money spent on town planning would have been better diverted into wage packets. The same year a twelve-acre athletic field was laid out. A six-story building opposite the plant housed the largest gymnasium of its time in Ohio, together with a fully equipped theater, class rooms, and club rooms. Company teams in most major sports were fully supported. When Goodyear expanded into southern United States in the 1930s, a similar model of paternalism followed. In the factories and cotton mills recreational facilities were provided as well as housing and town planning where needed, together with benefits such as the pension plan and the eight-hour day that Goodyear pioneered in the south.

From the time of the World War I, a comprehensive apprenticeship program existed to train high school graduates for technical careers. At the same time, acculturation classes were set up for immigrant employees all of whom were required to become citizens and speak proficient English. Consistent with the Goodyear belief in excellence and the athletic ideal, an elite core of workers was established and formed into "Flying Squads." These specially trained groups served as back-up in strikes, fires, and emergencies and as pioneers in setting up new plants.

The beliefs that healthier workers make for healthier, more productive organizations and that environmental engineering is the best way to prevent employee problems have survived until today. However, the creation of environments for and supposedly in the best interests of others is open to the same criticism now as it was at the height of industrial welfarism, namely, that mental health depends at least to some extent upon the participation of workers in the means of production and in decisions that affect their well-being.

The Case of English Quaker Companies

Although the companies created by Quakers varied to some degree in their approach to employee welfare, they were characterized by the practice of an ethic at the core of which was the imperative that all members be industrious, frugal, temperate, and honest. More succinctly, they were to remain "spiritual and solvent" (Windsor 1980).

This Christian code regulated the affairs of many early Quaker businesses to which their founders had turned as a result of being excluded from employment in higher education, the professions, and politics in seventeenth-century England. A surprising number of modern companies have Quaker roots. Among them are Lloyds Bank and Lloyds of London, Barclays Bank, the Great Northern Railway (a forerunner of British Rail), the London (Quaker) Lead Co., Clark Shoes, Truman and Harbury Breweries, Allen and Hanbury Pharmaceuticals, Huntley and Palmer Biscuits, Frys, Cadburys, and Rowntrees). In addition, Price Waterhouse, Wedgewood, and companies that formed the basis of Unilever, I. C. I., and British Steel all had Quaker roots. Because of their religious bonds and enforced intermarriage, the strong norms of the Quakers tended to prevail across these early companies, most of which were intially family owned and operated.

Quaker society is and was typified by an absence of hierarchy, consensus decision making, and accountability to the local congregation. Quaker businesses were characterized by innovative and much admired practices with regard to pricing, delivery, accounting, quality, and marketing. From the outset, value was placed upon the worker's quality of life because it was the Christian way and it made good business sense, usually in that order. Great store was placed on a practical education and employee access to continuation classes was a feature of many Quaker businesses; often this was facilitated through the Local Friends meeting house. It was each individual's duty to work hard and prosper, as well as to facilitate the welfare of others and not to prosper at their expense. Thus, during a period in which long hours, dangerous, unsanitary conditions, and low wages were the norm, Quaker companies became leaders in the improvement of the worker's lot. This took place in the context of a broad social welfare philosophy that promoted significant advances in penal reform, hospital care, education, housing, and the abolition of slavery.

Rowntree of York perhaps exemplifies the Quaker approach to industrial welfare during the late 1800s and early 1900s. The president, Joseph Rowntree, ironically, was a self-styled nonbeliever in philanthropy. He believed that if the structural problems of society could be alleviated there would be no need for charity. In this regard he resembled some of the other paternalists whom we have discussed. However, there was a difference in intent since Rowntree's provision of benefits such as housing in the form of the planned community of New Earswick was not to make workers dependent but rather, as he thought, to free them from the tyranny of squalor, high rents, and a generally adverse environment. Although New Earswick was adjacent to Rowntree's factory, its reasonably priced housing was not only for employees but was available to all working people. Similarly, the requirement that girls under the age of seventeen had to attend, on paid work time, classes in cooking, homemaking, home nursing, and dressmaking was predicated on the view that the scientific approach to domestic skills and hygiene would benefit the employees' own families as well as the families they would establish as married women after they left Rowntree. On-site education was made available on the premise that the family life or future prospects of young employees should not be impaired by their need to work for a living and if at all possible their potential should be enhanced. Thus, for boys under seventeen, continuation classes were provided in mathematics, English, and woodwork. In 1907 a full program of "Swedish physical training" courses was instituted. Both boys and girls were exposed to a wide range of social, athletic, and cultural experiences at their employer's expense.

Inflation-indexed pensions were introduced in 1904; the fund was inaugurated with a personal donation from Rowntree so that annuities could be immediately payable. Thereafter, employee contributions of 2½ percent to 5 percent were matched 150 percent by the employer. The Widows Benefit Fund was set up during World War I.

The Works Council was established in 1912 shortly after annual board meetings were opened up to workers as well as to stockholders (Rowntrees had become a limited liability company in 1897). The council comprised equal numbers of employee and management representatives. Its functions included revision of certain basic rules. For example, it was the council that decided to close on Saturdays and consequently to extend weekday hours, to provide one week's paid holiday (not nationally legislated until 1938), and to make the appointment of foremen dependent on workroom approval. In 1923 an Appeals Committee consisting of a chairman, two elected employee representatives, and two appointed management representatives was set up to review disciplinary actions. In the same year profit sharing was introduced at Rowntree's instigation after a long struggle with the board dating from 1906.

It was in this emerging context of working conditions that Rowntree introduced his first welfare worker in 1891. Her function was basically to look after employees' health and deal with behavioral problems. By 1892 an assistant had been hired and in 1904 more welfare workers were introduced in the wake of the first plant doctor whose identification of poor teeth as a major cause of illness led to the appointment of a company dentist. Rowntree said that healthful conditions of labor "are not luxuries to be adopted and dispensed with at will, they are conditions necessary for success. In keen international competition the vigor and intelligence of the workmen are likely to be a determining factor" (Windsor 1980, 148). As we have seen, this perceived link between health and productivity was not unique, although differences existed in the context within which the connection was made. In Ford's case, health was a commodity to be manipulated to his own ends, much as it was to the remote Board of Directors at Amoskeag. At Goodyear, the ecological balance of the firm depended on a healthy work force, individual members of which were rewarded for their commitment to the ideal. At Rowntree, however, health was not only instrumental to achievement of the company's goals, it was a condition of life that was perceived as essential to the dignity of individuals according to the Quaker tradition. In that regard, the promotion and maintenance of health was part and parcel of working conditions that set a value upon individual worth for its own sake and that were predicated on the belief that employment is more than the exchange of money for services.

The Cadburys

The crowning achievement of the Quaker Cadbury family was the founding of Bourneville in 1879, a new planned model community for employees of the new works. In establishing Bourneville and the new factory adjacent to it, the founders Richard and George Cadbury believed that obligations to employees did not end with wages: "We consider that our people spend the greatest part of their lives in their work, and we wish to make it less irksome by environing them with pleasant and wholesome sights, sounds and conditions" (Windsor 1980, 84). In 1900 the then 500-acre village was transferred to a trust. The gift was absolute and Cadbury surrendered all interest to capital and revenues. Stringent provisions were built into the trust so that the house to garden ratios remained one to four, that one-tenth of lands outside of roads and gardens were reserved for parks, and that factories and shops would never exceed one-fifteenth of the total acreage. Arrangements were put into effect to keep the public areas solvent and in repair. In its comprehensive attention to the needs of employees with regard to housing, sanitation, health, recreation, and conditions of work, Bourneville was an attempt to deal with the structural roots of social problems through socio-environmental engineering (Windsor 1980; Cadbury 1912). Indeed, there is some objective evidence that for a time, at least, the health benefits were substantial. During the period 1905–1910 the death rate per 1000 in Bourneville was reportedly 5.7 compared with 10.5 in neighboring Birmingham and 14.6 in England and Wales. Infant mortality was 62.4 per 1000 births for Bourneville, 87.6 for Birmingham, and 117.4 for England and Wales. Bright though these figures appear, it is unclear to what extent these differentials were affected by the preselection of a healthier work force, since a medical examination was compulsory for all job applicants. In addition, the work force as a whole was young.

At Cadbury, both academic and physical education was mandatory for employees between the ages of fourteen and eighteen. The technical and general education required for job preparation was much in evidence but so too were courses in social philosophy, political economy, and the like for those workers who were in low-level jobs and likely to remain so but whose private lives could be potentially enriched by exposure to a broad range of interests. Skills were also taught in this spirit as well as with the belief that domestic and handicraft courses would prepare workers for life in general, not just for the factory. All employees could avail themselves of the classes taught in the evenings. Fees were charged to volunteer students but were refundable on 85 percent attendance.

A comprehensive physical education program appears to have existed from Bourneville's inception with resources such as a covered and heated women's pool, an outdoor men's pool, playing fields, and gymnasia complete with staff. Courses for the fourteen to eighteen age group were offered during work hours, 2½ hours per week. Voluntary classes after hours were open to all employees. They covered, among other things, calisthenics, aerobics, swimming, life saving, dancing, and weight training. In addition, athletics clubs run by employees made full use of the facilities during off-work hours. In all, the women's and men's athletic facilities covered twelve acres each. The Cadburys seem to have been well aware, however, that these benefits amounted to no more than frills if the conditions of the work itself were poor. Edward Cadbury said, "Clubs and classes, saving funds, libraries etc. do not compensate the worker for low wages, long hours and unsatisfactory conditions" (Cadbury 1912, 263).

Cadbury appears not to have had welfare officers or secretaries distinct from the medical department, which by 1906, through its two doctors (one man, one woman) and four nurses served various social work functions. Nurses and doctors visited the sick or injured at home. Nurses also made reports on the progress of the convalescent to the medical department and to the department forewoman or foreman and continued to follow up on employees when they returned to work. On the basis of their observations the nurses might instruct supervisors about light duties for convalescents, provide supplementary food vouchers, or send patients, for two to four weeks, to a convalescent home that the firm operated. A stay in these homes was sometimes used as a temporary breather for employees (particularly girls) in adverse family situations that were affecting their physical and mental health. In addition to performing these rehabilitative duties, the nurses visited homes systematically, advising on hygiene, nutrition, ventilation, and other public health matters.

From its earliest days and very much in keeping with Quaker tradition, the Cadburys emphasized record keeping as the framework for their disciplinary system. They used warnings followed by suspensions and dismissals but records were destroyed after two years in cases where the worker's performance improved. Unsatisfactory workers who were slow or often late were frequently found to be either in poor health or to be poorly suited to the jobs they were in. In the case of the former a variety of medical and social welfare options were used including sick leave, convalescent homes, supplemented diets and so on. In the case of the latter, employees would be moved around until the best fit was achieved between themselves and jobs they could perform adequately. The use of such measures was reported to have reduced the use of disciplinary measures from 229 in 1889 to 70 in 1910 despite increases in the size of the work force. It is not clear whether other factors contributed to this reduction but it is evident that health-related interventions and transfers based upon ergonomic considerations played a significant role in diverting workers from the disciplinary process.

Cadbury represents one of the most integrated and comprehensive implementations of the Quaker industrial ideal. For modern EAP and HPP observers, the Bourneville experiment is instructive indeed, incorporating as it did both structural and programmatic strategies for the promotion, maintenance, and restoration of workers' physical, mental, and spiritual health. It is only with great difficulty that today's corporations manage to develop anything like the integrated vision of the healthy, satisfied, and productive worker that guided the Cadbury enterprise. Yet, for all of its comprehensiveness, the approach was still paternalistic in that the social welfare package that was offered the worker was defined by the employer and assumed what workers needed. Even so, the philosophy of the Cadburys, as with Rowntree, was that the quasi-prosthetic structure of planned community, benefits, and welfare would liberate rather than enslave, foster independence rather than dependence. Indeed, employees were encouraged to participate in the management of all areas of their work life since control was largely by committee. Employee representatives often constituted half the membership of joint management—worker committees. Sometimes, especially in recreational matters, the committees were entirely made up of workers. There were standing committees on the pension fund, education, grievances, one that judged suggestions for improving operations, another that regularly inspected the sport facilities, and a great many more. However, the assumption remained that if the value of the broad benefits had been converted into larger wage packets, workers would not have been able to achieve their own financial independence, secure their own futures, or remain healthy.

Reflections of the Past in Modern EAPs and HPPs

The purpose of the historical excursion that we have just been on is to illustrate the fact that EAP and HPP concepts are rooted firmly in a paternalistic tradition that can be traced back to the earliest days of employment relationships and indeed even to feudal and prefeudal times. These origins, we contend, have left their marks on the modern manifestations of employer welfarism as expressed in the form of employee assistance and health promotion. Some accounts of the history of EAPs relate it principally though not wholly to the growth of industrial alcoholism programs (e.g., Trice and Schonbrunn 1981; Trice and Beyer 1984). Indeed the reference to job-based alcoholism programs, which emerged during the 1940s, as the "immediate model" for

EAPs (Trice and Beyer 1984, 251) is no doubt accurate. Before the development of these programs, it appears that even in the most benevolent of paternalist regimes, the excessive use of alcohol tended to be dealt with in a cavalier fashion, often resulting in outright dismissal. A possible explanation for this is that employers of the highly paternalistic ilk may have believed that alcoholism *should not* have occurred in the healthful conditions that they had created for employees, and was an indication of the individual's moral failing. Whether this attitude became more marked as the benevolence of the employer in other respects increased is a matter for conjecture. In any event, it was in the context of Industrial Alcoholism Programs (IAPs) as they evolved that standards for the official management of employees with alcohol problems emerged. The eventual insistence that case finding be carried out on the basis of deteriorating job performance rather than in relation to the identifier's idiosyncratic concept of alcoholism was a major step forward in professionalizing the management of alcoholism. It was accompanied by a recognition that supervisors and managers would need to be trained in documentation, warnings, and the whole process of constructive confrontation. Nonetheless, while we do not dispute the line of descent from IAPs to EAPs and consider it to be an important genealogy, we believe that both broad-based and alcoholism-specific programs need to be located in a more general historical context in order to identify what is wrong with these programs today.

The proposition that something is wrong with EAPs and HPPs is, of course, contentious. To some extent, however, this case has been made already in the matter of EAPs (see Shain and Groeneveld 1980; Walker and Shain 1983; Shain 1985). The gist of the problem is that most EAPs fail to identify a large proportion of the population at risk either in terms of alcohol-related problems or in terms of the wide range of other problems with which they are meant to deal. Once identified and referred, the available evidence suggests that employees with problems are likely to do fairly well as long as we accept what are primarily management-oriented criteria for success. Chapter 7 is a reprise of these issues in the context of how some of these limitations may be overcome. However, it should surprise no one that these limitations exist because essentially they were preordained by their origins, which can be found deep in the folds of the paternalistic cloak that still shrouds a large part of the workplace.

The situation with HPPs is not much different. As programmatic entities, health promotion initiatives have gained a new lease of life in the late 1970s and 1980s (for a typical review, see Parkinson et al. 1982). However, as our sample of historical case studies illustrates, they are really just a modern expression of an old idea. As chapters 2 and 7 among others will hopefully demonstrate, HPPs bear the indelible brand mark of employer paternalism. Too frequently, the programs are initiated on behalf of employees without their consultation and too often communicate the view that deteriorating health is a moral failing for which the individual worker is responsible.

In short, by default or design, EAP and HPP practitioners often become aligned more with employer interests than with employee interests. If, as this book contends, EAPs and HPPs are to grow closer in the future, perhaps becoming welded into comprehensive approaches to worker health and welfare, this issue of agency will become even more acute. One can envision the specter of a beautifully coordinated employee health and assistance program (such as chapter 7 outlines) that is so slanted toward the employer's interests that it becomes simply a bigger and better way of squeezing the last ounce out of workers. Thus, if we are to move toward a greater coordination of effort between EAPs and HPPs it will become more important for practitioners to test their practices against the question, "Whose interests are we serving?" The somewhat dramatic examples from the past act as reminders of the widely varying contexts in which the goal of the healthier worker may be pursued. From these examples it is clear that a value may be placed upon health for a number of quite disparate reasons: Health can be a commodity in a calculative, "canned welfare" approach to improve productivity; it can be part of an individual employer's ideology in which health is a condition for the efficient operation of the human machine; or it can be an element in a perceived ecology of individual, organization, and society. Health may be defined in purely physical terms or it may be seen as a body–mind condition that needs to be nourished in the context of working environments that facilitate the self-respect and dignity of workers. As suggested earlier, all these conceptions of health are to one degree or another projected into the fields of modern EAPs and HPPs. The challenge for practitioners is to recognize with what concept of health they are aligned and to seek realignment if a dissonance is perceived between their professional standards and what they actually do.

The material that follows examines these issues more closely, attempting to bring out implications for the roles of EAP and HPP practitioners in the framework of the over-arching question: Who is served? Chapter 2 examines health promotion from this perspective and suggests how practitioners might design programs to be more in line with employee needs and consequently to be more effective. Chapter 3 demonstrates, through research conducted by the authors, the extent of these employee needs, while chapter 4 is an overview of the kinds of response that currently exist to these needs. Chapters 5 and 6 are accounts of evaluations conducted by the authors upon relatively new health promotion approaches, with implications for future modifications to bring them more in line with employee interests. Chapter 7 looks at the potential synergy of coordinating EAPs and HPPs. In Chapter 8, the impact of the work environment on mental and physical health is reexamined with a view to suggesting roles that EAP and HPP practitioners can play in the modification of structural factors associated with the organization of work. Chapter 9 pursues this topic into the somewhat misty interface area between programmatic and structural interventions, again suggesting extended roles for practitioners of EAPs and HPPs, particularly as they relate to the development of social support for health in the workplace. Much of what we have to say in the body of this book is directed, in the context of EAPs and HPPs toward the creation of a balance between employee and employer interests since it is our belief that the scales have been historically tipped in favor of the latter. This is not to suggest that employees have not been helped in the past. Clearly they have been and much credit is due to the dedicated people who staff the programs responsible for this success. Our point is that much more can be done. In this regard, we feel that we do no more than reflect the frustration that is so often expressed when EAP and HPP professionals gather (usually under separate auspices) to discuss their work. In this book, we set out to identify some of the dimly perceived origins of this frustration, suggesting that only by the adoption of new and extended roles will EAP and HPP professionals free themselves from the shackles of the past that keep them on the plateau of which they complain. As noted in the introduction, this book is written out of respect for their desire to achieve the lofty goals that the standards of their existing and emerging professions define for them.

We do not suggest that there is any absolute resolution of the agency question. In order to serve employees, it is obvious that employers must too be served. The challenge is how to accommodate the needs of both without the one being done at the expense of the other.

2

The Relationship between Current Life Style and Future Wellness: Implications for the Design of Health Promotion Programs

D espite the worker resentment that followed in the wake of imposed health promotion, the validity of the relationship between life style and health was not really at issue. The exact nature of this association and of the variables that affect it, however, was not considered seriously by scientists until the second and third quarters of this century. From crude beginnings, a body of rudimentary principles has emerged in the last quarter of the century that describes how life style and health are related and regulated. This emerging body of knowledge serves as a restated rationale for the development of health promotion in the workplace according to certain principles of effectiveness. This chapter examines the evidence and its implications for the work of modern health promotion professionals. While much of the research has been done outside the workplace proper, it has usually involved employed people and is certainly of relevance to the management of industrial programs.

Overview

The popular wisdom that linked moderation in all things to a long and productive life has been vindicated and elaborated upon by a now substantial body of research. While it seems clear that current life style does in fact influence future health in a highly significant manner, our review suggests strongly that the interactions and interdependencies between behaviors related to eating, drinking, sleeping, smoking, recreation, exercise, working, and coping with life in general are better predictors of morbidity and mortality than any single behavior in the constellation. Although estimates of risk as related to future morbidity and mortality are still fairly gross they have developed considerably from not so long ago when risk factors were seen primarily through a medical lens and consisted largely of discrete "health practices." We trace the development of conceptualizations about risk factors from this early stage through to the gradual acceptance of a more holistic view of health and well-being.

In the process of development, the idea of health-related behavior as "discretionary"—that is, behavior under the control of the individual—is being eroded by the results of studies that have demonstrated the importance of socioeconomic, cultural, and political factors in the genesis and maintenance of personal health. The concept of life style appears to have marked an era of transition from a time when health was seen as the by-product of several specific behaviors to a time when environmental stressors are seen as major influences upon the adoption of healthy practices. The term life style itself, though still popular, may fall into disrepute as it becomes increasingly clear that its implication of a "chosen way" is often not grounded in reality. Scrutiny of this implication has been substantially sharpened by investigators who have gone a long way toward demonstrating the importance to health of such basic social factors as gender identity and socioeconomic class. Furthermore, the possibility that some behaviors that are supposedly maladaptive from a health point of view may serve socially adaptive, "coping" functions—for example type A or coronary prone behavior (CPB)—adds to an emerging perspective that sees health as a result of how we function or are caused to function in society. By the same token, life style often emerges less a chosen way of living than a mould into which we are poured.

In this context, individual responsibility for health remains a necessity but it is a form of self-defense that will yield the greatest gains only in situations where collective responsibility supports and helps to generate the will to thrive. In so far as holistic views of health portray the individual as functioning in constant interaction with the social environment, it follows that the health of individuals must in some measure depend on how that society functions. At a more manageable level, health is influenced by the manner in which organizations function (see also chapter 8). In this context, the role of occupational stress is often examined, although there is still a lamentable tendency to ascribe the source of this stress exclusively to individual weakness rather than to indict collective insensitivity as an equally culpable partner.

Research has tended to confirm the symbiotic nature of the relationship between mental and physical health—a relationship informally postulated for as long as our social memory serves. The task remains, however, to convert this wisdom into a form of social intervention that reflects this inseparable union of mind and body. Current health promotion programming is rarely designed in such a way as to even recognize this union, let alone reinforce it. Fragmentary "rational" treatment of isolated risk factors is still the order of the day rather than a holistic approach to the individual in society. Research methods are correspondingly underdeveloped, so that pleas for multivariate analyses of health behaviors and of changes wrought in them still appear rarely and are even less often carried out.

The success of health promotion approaches aimed at improving well-being and extending longevity will depend upon the incorporation of certain principles into their construction, whether the programs are intended for the community, the workplace, or both. First, the relationship between health-related behavior and a wide variety of social, cultural, and economic factors must be acknowledged, the corresponding interdependence of eating, sleeping, exercise, work, and recreation patterns recognized. Second, a corollary of the first, it must be accepted that environmental, organizational, and social support for change and reinforcement of health-related behavior is essential. Third, the importance of introducing affective components into health promotion programs must be recognized so that individuals feel competent to undertake change and feel better about themselves for doing it.

The underlying principle is postulated to be that individual health-related behaviors not be approached in a "health vacuum." We must be prepared to examine and modify the social, cultural, economic, and organizational supports for unhealthy behavior. This is not an unattainable goal. Programs of even a short duration can be designed in such a way as to help people clarify for themselves the extent to which their behavior or "life style" is constrained by external influences and to help them find chinks in the walls that appear to surround them.

Context of Concern About Present Life Style and Future Health

An increasing acceptance of a holistic definition of health has emerged as a recurrent theme in the literature of the postwar era (Rosen 1949; Dunn 1959; Porterfield 1960; Freeman 1960), and often drew support from the developing research on the role of stress in health and illness (Selye 1950, 1956; Sanua 1960). In 1960 the *Journal of Health and Human Behavior* was established as a forum for debate and research in the developing field of "social medicine" (Porterfield 1960).

The emergent need for a more generalized approach to health ("beyond germ theory" [Galdston 1954]) paralleled increased state expenditures on health and the rise of what has been called the welfare state (Finkel 1977; Navarro 1976; Walters 1982). The twenty-five year period from 1945 to 1970 saw marked increases in government social expenditures, which included the establishment of the National Health Service in Britain (1948), the introduction of national health insurance in Canada (Bill C-277, 1968), and increased overall spending on social welfare services ("the welfare explosion," Wilson 1977). According to Roemer (1984) in the last thirty years industrialized countries have seen health as a proportion of GNP increase from 3 to 5 percent to 5 to 10 percent. Developing countries have experienced increases from 1 to 3 percent to 3 to 5 percent. Navarro also points out that in the United States, France, England and Wales, and West Germany health expenditures, as percentage of the GNP, showed steady increases from 1950 to 1973. In the United States, for example, it went from 4.6 percent in 1950 to 7.7 percent in 1973. Correspondingly, about 7 percent of the average U.S. auto worker's wage package of approximately $30,000 in 1980 consisted of health benefits (Sapolsky et al. 1981). Crawford (1977) cites former U. S. President Jimmy Carter as reporting that "the average American worker is now devoting one month's worth of his yearly salary just to pay for medical care costs" (p. 665). Parkinson and associates note that U. S. total medical care expenditures were 9.4 percent of GNP in 1980, and are expected to rise to 9.9 percent of GNP in 1985 (Parkinson et al. 1982, 1).

Empirical Evidence of the Relationship between Life Style and Future Health

Health Practices and Risk Factors: Epidemiology

The development of the concepts of life style and a holistic definition of health can be seen in the studies of the Human Population Laboratory (HPL), Alameda County, California, which attempted to quantify the World Health Organization's (WHO) 1948 concept of health—a state of "physical, mental and social well-being" (Belloc and Breslow 1972). In an early study from HPL's 1965 survey of 6,928 adults in Alameda County, Belloc, Breslow, and Hochstim (1971) attempted to assess population health levels generically rather than in disease categories. In their exploration of the relationship of health to economic and social variables, health was conceptualized as a spectrum with three axes; mental, physical, and social. It was hypothesized that an individual would have related positions on all three scales, and "ways of living" would affect and be affected by each realm.

Participants' physical health was classified along a spectrum of health that emerged from individual health reports categorized as follows:

Category 1 (severely disabled) included those who reported trouble with everyday activities such as feeding and dressing, or who were unable to work for six months or longer.

Category 2 included those who had a slighter disability and had changed hours or type of work or had restricted some of their activities.

Categories 3 and 4 included those reporting one chronic condition in the past year.

Category 5 reported one symptom last year but no disabilities or chronic conditions.

Categories 6 and 7 had no complaints but reported varying energy levels.

Men were found to have fewer symptoms and disabilities than women, higher energy levels, but the same number of chronic conditions. Chronic conditions and disability, as expected, increased with age. In the age group 20–24 the proportion of severely disabled was 1 percent while in the age group 75 and over it rose to 30 percent. Part of this age differential, however, was found to be attributable to the influence of lower income and education among older people. In relation to ethnicity, Japanese and, to a lesser extent, Chinese people showed more favorable health than blacks or whites, with little difference between blacks and whites.

Belloc and Breslow (1972), using the same data base (the 1965 HPL survey) continued the attempt to arrive at a general indicator of health according to which populations could be readily compared and upon which the influence of personal habits could be measured. Participants' physical health status was again categorized according to the spectrum of health just outlined. The categories were then correlated with personal habits—cigarette smoking, alcohol consumption, sleeping habits, physical recreational activities, and nutrition (measured by regularity of meals and weight in relation to height).

Taking into account the finding of the previous study (Belloc, Breslow, and Hochstim 1971) that age is inversely related to good health, results here showed that (with the exception of persons over 75 years of age) as the number of good health habits rose so did evidence of better health, *across* age groups. Further, the average health status of those aged 35–44 who followed fewer than three of the "good" practices was similar to those over 75 who followed all of them; those aged 55–64 who followed all of the good practices showed physical health status identical to those aged 25–34. There is some difficulty in comparisons between age groups, however, since particularly among older people, we are seeing only the survivors. Nonetheless, this suggests that the survivors are those older persons who have maintained good health habits.

In 1974 HPL carried out a follow-up study of 96 percent of the 1965 sample for mortality, and was able to ascertain by questionnaire the 1974 health status of 4,864 (70 percent of the original sample) of the survivors. In relation to this sample, Breslow and Enstrom (1980) explored the relationship between seven life-style variables and health and mortality. The seven personal health practices (physical exercise, never smoking, abstinence or moderate use of alcohol, seven to eight hours of sleep regularly, proper weight maintenance, not eating between meals, and eating breakfast) were accumulated to form health practice scores of 0–3 (poor), 4–5 (fair), and 6–7 (good).

Results showed a strong relationship between health practice scores and both longevity and subsequent health status. Men following 7 health practices had 28 percent of the mortality rate for men following 0–3 practices. Women following 7 health practices had 48 percent of the mortality rate for women following 0–3 health practices. Health practices were relatively stable over time. Of men reporting 0–3 health practices in 1965, 72 percent were at the same level in 1974; only 7 percent had changed from "poor" to "good." Of men reporting "good" health practices in 1965, 60 percent were at the same level in 1974; 27 percent had dropped to 5 health practices; and only 3 percent had dropped to 0–3 health practices. Similar data were apparently found for women. The average number of health practices for men and women in both 1965 and 1974 was 4.9. Over the nine year study period, trends toward both obesity and being underweight (both associated with higher mortality risk) and toward a decline in smoking, reflected trends in the United States as a whole. There was also a trend away from eating breakfast every day.

The possibility of undetected disease, illness, or impending death at the time of the 1965 survey raises the question of direction of influence: Did poor health habits influence sickness or vice versa? For example, those who were already ill in 1965 may have reported not eating breakfast or sleeping well. If this were the case, a high proportion of the excess mortality rate would be expected to emerge within two and one-half to five years, and not show up much after that. In fact, among those with 0–3 health habits mortality was 250 percent of the *expected* rate for the population at large in the first two and one-half years (accounting for the fact that very sick persons probably did not participate in 1965). This held for both sexes.

In the final four years of study (five and one-half years after the effects of impending death should have ended) men with 6–7 health practices still had 50 percent of the mortality rate for men with 0–3 health practices. Over the entire nine and one-half years women with 6–7 health practices had 50 percent of the mortality rate for women with 0–3 health practices, and 60 percent during the final four years. For all three levels of health practices differences in mortality rates were significant. Those with 6–7 health habits showed fewer deaths by cancer and cardiovascular disease and fewer chronic conditions and disablements. Age-adjusted mortality rates were adversely related to good health practices, with "moderate" alcohol consumption more closely related to low mortality rates than "no" alcohol consumption.

Breslow and Enstrom's general conclusion that "following good health practices favors longevity" (p. 483) is also supported in Wiley and Camacho's (1980) study from the same HPL research. Wiley and Camacho appear to be the first of the HPL researchers to use the term "life style," which "in this context, consists of discretionary activities which are a regular part of an individual's daily living pattern" (p. 1). "Life style," then, refers to those aspects of life over which, supposedly, one exercises choice and control.

Participants (3,892 white respondents, under 70 years of age in 1965, who also responded to the 1974 questionnaire) were given physical health status scores along the seven-measure spectrum of disability, chronic conditions, symptoms, and complaints. The 1974 health scores (adjusted for age, sex, and 1965 health level) were examined in relation to the same seven health practices used in the previously discussed studies, with the following results:

1. and 2. *Eating breakfast and not eating between meals.* Contrary to Belloc and Breslow (1972), Wiley and Camacho did not find a general relationship to good health. The relationship did hold for men, but not for women, or for the sample as a whole.

3. *Cigarette smoking.* The most favorable scores were found among those who had never smoked. Light to moderate smokers who had quit smoking by 1965 showed the highest adjusted 1974 scores and the greatest improvement in health. Scores of former and current smokers were less favorable. The strongest association was in the male sample.

4. *Alcohol consumption.* Findings were similar to those of Breslow and Enstrom (1980) for mortality. Moderate alcohol consumption in 1965 was associated with the most favorable adjusted (1974) health scores. Abstinence due to ill health, however, was not accounted for.

5. *Leisure time activity.* There was a "clear and significant relationship between activity level and health score," even excluding persons who were disabled in 1965 scores. Most extreme differences were "between those who reported no activity and those who reported even a little activity" (p. 9).

6. *Sleep patterns.* Concurrent with Belloc and Breslow, findings were that less than seven hours was apparently a significant risk to health for both sexes.

7. *Weight in relation to height.* Extreme obesity was found to be a clear risk factor for men and women, but especially for men. Those who were 10 percent underweight also had a higher mortality risk, which appeared to be partly "due . . . to prior disease or disability" (p. 14).

A later analysis of the same data (Berkman and Breslow 1984) also accounted for social networks in future health status. Four kinds of relationships were considered: marriage, close friends and relatives, church membership, and associations with secular organizations. Briefly, "people with social relationships had lower mortality rates than people without them" (p. 49). "The more intimate ties of marriage and contact with close friends and relatives were stronger predictors of lower mortality rates" (p. 149). The trend was a general one however and "only in the presence of severe social disconnection . . . did mortality rates rise sharply" (p. 149).

While the HPL research has provided valuable insights into the relationship of life style to future health, there are notable gaps in the information. No data were collected on occupational health, for example. However, if life style and health practices are viewed, as Wiley and Camacho suggest, as "discretionary" it is perhaps understandable that no data were collected on working conditions and work life. While issues related to occupational health may be important, they may not always be "discretionary" matters of individual choice. This of course begs the question whether "life style" is as discretionary as the investigators in the HPL studies implied. In addition, given the widespread introduction of the birth control pill in 1963, and its linkage to serious health problems for women, including heart disease, one might suggest that female health information would have been more complete had data on contraception been collected.

In a sixteen-year follow-up study of children's health responses, Mechanic and Cleary (1980) developed an index of health behavior based on variables similar to those of the HPL research. Components of the index included taking risks (while driving), drinking, smoking, exercise, preventive medical care, and wearing seatbelts. General orientation to health and medical care was assessed by: drug-culture orientation, concern with health, feeling comfortable going to the doctor, having faith in doctors, and having a specific doctor or clinic. The variables most substantially related to the index of positive health behavior were being female and having more education (see chapters 4, 5 and 6 for elaboration of sex differences in health orientation). Poor health seemed to be linked to poor psychological health, but the dynamics of that relationship were not explored. Poor health was also related to less seatbelt use, greater likelihood of smoking, less preventive medical care, and less physical activity or exercise. We note, in passing, that correlational studies of this type should not be read as implying causality.

Approaching the issue from a somewhat different perspective, Zook and Moore (1980) investigated the characteristics of high-cost users of medical care in hospital settings. A review of 2,238 patient records in six contrasting hospitals yielded profiles of high-cost patients. In the five general hospitals, 13 percent of patients accounted for as much hospital billing as did the other 87 percent. Profiles of these high-cost users by age, sex, and "potentially harmful personal habits" revealed that 40 percent were over 65 years of age (compared with 15 percent of low-cost users), but sex differences were insignificant.

"Potentially harmful personal habits" included "substantial continued consumption of alcohol, heavy smoking, use of hard drugs, obesity, or persistent refusal to follow physician's advice." From 31 to 61 percent of high-cost patients had an associated unhealthy habit compared with 20 to 45 percent of low-cost patients, a difference significant at the .05 level. "Records of high-cost patients showed a pairing of diagnoses that are commonly associated, such as alcoholism and benign inflammatory disease" (p. 999).

Twelve percent of the high-cost upper quintile patients were admitted to hospital as a result of unexpected complications *due to treatment,* however (as opposed to 5 percent of the four lower quintiles). Unexpected complications during hospital treatment occurred a full thirteen times as often for the high-cost upper quintile patients. Fault for and controllability of these complications were not accounted for by Zook and Moore.

Personal habits and health practices as life style variables can also be conceptualized as risk factors in the development of disease. This approach is somewhat more disease-oriented, however, since according to this model health is measured by the absence of disease, rather than the presence of wellness. Although Zook and Moore's focus on personal habits and health places their study in the same general framework as Mechanic and Cleary's and the HPL research, it also spans the conceptual gap between defining lifestyle activities as predictors of future health status and defining them as risks for disease and death.

Risk factors lend themselves to intervention, and to the measurement of the effects of intervention on morbidity and mortality. While it is difficult to isolate specific causes and predictors of disease, disease risks can be identified and occur in clusters; for example, smoking cigarettes, high blood pressure, and high serum cholesterol levels are associated with heart disease.

In an effort to develop guidelines for health education, Kok and associates (1982) investigated whether there is "an underlying risk-taking way of life" for coronary heart disease (CHD) (p. 986). In a multistage, stratified random sample of the Dutch population, personal at-home interviews were completed for 1,951 participants, aged 18 to 64. Risk was measured in the areas of smoking, nutrition, obesity, physical activity, and "preventive orientation." No clear coronary risk life style was established, but "independent demographic and socioeconomic determinants of a 'risky' lifestyle were sex (male), education (low), and occupation (low)" (p. 989). Marital status and familial social class showed no association. Participants also had a distorted perception of the healthfulness of their own life styles; of those in the high-risk category (10 percent of participants), 55 percent believed their own life style was healthful. Because this study was based on self-reported data, however, the possibility exists that many participants reported only socially acceptable behavior to researchers. Kok and associates conclude, in fact, that health education should focus on attitudes rather than on knowledge.

In another Dutch study of risk indicators for CHD, Kromhout, Bosscheiter, and de Lezenne Coulandes (1982) surveyed 871 middle-aged men. Dietary information was collected in 1960, with a ten-year follow-up for mortality. Mortality from CHD was about four times higher for men in the lowest quintile of dietary-fiber than for those in the highest quintile, although the relationships did not hold after multivariate analyses. However, "rates of death from *cancer* and from all causes were about three times higher for men in the lowest quintile of dietary fiber intake than for those in the highest quintile, and these relations persisted after multivariate analyses" (p. 518).

Another risk component of life style strongly related to mortality and the incidence of disease is alcohol consumption. Moderate alcohol consumption has been shown to be a better predictor of low morbidity and mortality than abstinence. In a ten-year follow-up study of 1,422 British men, Marmot and associates (1981) related daily alcohol consumption patterns to all causes of death, death by cardiovascular disease (CVD), and non-CVD causes. In the four levels of alcohol consumption (abstainers, light, moderate, and heavy drinkers), the lowest rates of CVD mortality were for light drinkers. It was also found that light drinkers maintained the lowest mortality rates regardless of smoking status. Ex-smokers showed mortality rates similar to those of heavy drinkers.

Larbi and associates (1984) suggest a curvilinear relationship between blood pressure and alcohol intake. They analyze data from a 1950s' study of 11,899 middle-aged employed men and conclude that heavy drinking contributed to from 4.3 to 11.6 percent of hypertension, depending on cut-off points. They suggest that with steadily increasing per capita consumption of alcohol since the 1950s the population at risk for hypertension due to heavy drinking is also increasing.

✳ Risk Factors: Intervention

The main cause of mortality in industralized countries is cardiovascular disease (CVD), especially for middle-aged men. Important risk factors for CVD are smoking cigarettes, high plasma-cholesterol concentrations, and high blood pressure. Several community-based studies have been carried out to determine the possibility of reducing CVD risk and incidence through health education. These studies suggest principles that may be applied to workplace interventions (program examples are given in this chapter and also in chapter 7).

Researchers from Stanford University carried out a field experiment in modification of risk factors in northern California (Farquhar et al. 1977). In an attempt to overcome problems encountered in previous attempts to change diet and smoking habits, the study used a mass-media campaign combined with face-to-face instruction. The mass-media materials not only offered information but also were designed to teach behavioral skills and to influence motivation and attitudes. Self-control training and behavior-change methods were incorporated into both mass-media and face-to-face approaches.

In three comparable communities, data on coronary heart disease (CHD) knowledge, daily cholesterol intake, smoking rate, blood pressure, cholesterol concentrations, and relative body weight were gathered in annual interviews for two years from over 1,200 men and women aged 30 to 59. The control community (Tracy) received no mass-media campaign or counseling. The two other communities (Gilroy and Watsonville) both received mass-media campaigns, and an additional two-thirds of Watsonville CHD high-risk participants were randomly selected for additional counseling.

From "remarkably uniform" baseline values, after two years, effects of the media and the media plus face-to-face counseling campaigns had "significant positive results on all variables except relative weight" (p. 1194). Changes in knowledge and risk factors in Gilroy and Watsonville were not only maintained after the first year but further increased in the second year. Those who received face-to-face counseling among the high-risk Watsonville participants showed a greater decrease in risk factors at the end of the first year than the media-only groups (p. 1194). However the difference between the two approaches was reduced in the second year as the media-only groups showed further gains.

Farquhar and associates conclude that in general changes in knowledge, behavior, and physiological endpoints were maintained, and even improved. Intensive individual counseling seemed important for changing behavior such as cigarette smoking and diet. The cost-effectiveness of mass-media education appeared to outweigh that of face-to-face instruction. There was, however, no control for socioeconomic status in analyzing the effects of intervention.

In the 1970s World Health Organization statistics showed considerable differences in CVD rates among industrialized countries. According to figures quoted by Salonen, Puska, and Mustaniemi (1979) the lowest age-standardized mortality from heart disease of middle-aged men was found in Japan, the highest in Finland (p. 1178). From 1950 to 1973 most countries showed decreases in CHD mortality rates, while in Finland CVD mortality and regional differences in mortality increased in the 1960s, but national rates leveled off in the 1970s. The highest CVD rates were found in North Karelia, a rural county in eastern Finland.

The high CVD rates in North Karelia prompted a petition from the local population requesting the government to take action leading to the establishment of the North Karelia Project in 1972. The North Karelia Project was a comprehensive community program intended to improve services and change both behaviors and environments, in an effort to improve detection and control of hypertension, reduce smoking, and promote low cholesterol diets (McAlister et al. 1982). It was not an experiment to test the relationships of life style, risk factors, and CVD; rather, it was a community project specifically aimed at lowering CVD rates.

✗ The program focused on primary and secondary prevention, working within the existing community health and social services. McAlister and associates outlined the planning and implementation, which were carried out within the following areas.

✗ 1. *Improved preventive services to identify high-risk persons through detection and treatment of hypertension.* The approach was to change existing services, rather than to promote their use. Changes included increased responsibility for public health nurses, mass screening programs for hypertension at county fairs and village centers, public health nurse referrals for and surveillance of those with elevated blood pressure, and regular mailings of reminders of follow-up visits to all persons recorded in the public health nurse's hypertension register.

In 1972 the baseline survey showed that 13 percent of male hypertensives in North Karelia and Kuopio (the neighboring county, which served as a reference area) were receiving anti-hypertensive treatment. By 1977, 45 percent of male hypertensives in North Karelia were being treated and 33 percent in Kuopio (where new services were being modeled after those in North Karelia).

✗ 2. *Information to educate the community about health maintenance and disease prevention.* Applying communications theory and research, complex information was presented as news, in several steps, simply and frequently for maximum comprehension and retention. From 1972 to 1977 over 1,500 articles on CVD risk factors appeared in local newspapers, and over one-half million leaflets, posters, signs, and stickers were distributed. Local groups were contacted to distribute material and organize health education meetings and 251 general meetings of local organizations reached over 20,000 community members.

Baseline data showed differences in knowledge between different educational and occupational groups, but all showed 10 to 15 percent increases in knowledge, awareness, and understanding of CVD. Similar changes in Kuopio were attributed to the effects of heightened national media attention in both counties.

✗ 3. *Persuasion to motivate environmental and life style change in the direction of good health.* The project sought to combine credibility, emotionality, and commitment to behavioral change in its approach to persuasion. Messages were disseminated by various sources, from the World Health Organization to local formal and informal opinion leaders with content anticipating and answering counter-arguments. Messages that might provoke fear were tempered by "clear and attainable recommendations for reducing . . . fear." Surveys showed that North Karelia decision makers had received much more dietary information, and there were significantly greater reductions in fat intake than in Kuopio. No tendencies of increased anxiety or psychosomatic complaints were exhibited. On the emotional level the project associated its goals with the pride and provincial identity of the community. Participation and changes were urged not only for individual health benefits but for all of North Karelia. "For instance, signs reading 'Do not smoke here—we are in the North Karelia Project' were everywhere and fostered a kind of local patriotism" (p. 46).

4. *Training to increase skills of self-control, environmental management, and social action* was implemented in four steps: modeling, guided and independent practice, feedback, and reinforcement. Recognizing that most North Karelian women occupied traditional roles, the project worked with the local housewives' association to introduce "parties for long life," teaching healthy food preparation skills to 15,000 participants at 344 sessions (p. 46). In each village housewives gathered for demonstration, participation, and guidance in cooking healthy meals. Their families were then invited to enjoy the meal with them in the evening, naturally reinforcing the cooking and nutrition lessons. Participants were also shown that cost was lower than in the traditional methods. In 1976, 18 percent of women and 9 percent of men had been involved, and the "parties" were adopted nationally.

5. *Community organization for social support* was encouraged through weekend training programs for local leaders. These natural leaders were urged to see themselves as models and set positive examples for the community. Although over 1,000 local leaders were involved, Puska and associates (1981) reported no personal initiation of CVD-control measures on their part but greater satisfaction with CVD-control activities on the part of local health professionals.

6. *Environmental change to create opportunity for healthy actions* was achieved through increased availability of low-fat foodstuffs and a reduction of the proportion of people drinking high-fat milk by 40 percent. Voluntary restraint on tobacco promotion in North Karelia was reflected in national law in 1977.

Significant reductions in risk estimates of serum cholesterol, smoking, and blood pressure in North Karelia are reported by McAlister and associates (1982) and Puska and associates (1979). In general, trends were reversed between North Karelia and Kuopio from 1972 to 1977, but risk reductions also occurred in Kuopio, and are attributed to a spillover effect.

Salonen, Puska, and Mustaniemi (1979) reported changes in North Karelia mortality and incidence of CVD, monitored through registers of acute myocardial infarction (AMI) and stroke and data on death certificates. Statistical *decreases* in incidence and mortality rates in North Karelia for the duration of the project were: in total mortality, 5 percent; in CVD mortality, 13 percent for men, 31 percent for women aged 30 to 64 years; in AMI incidence, 16 percent for men, 5 percent for women; with respect to cerebral stroke, 38 percent for men and 50 percent for women.

The difference in mortality changes between North Karelia and Kuopio were not significant, but a "comparison of the changes in the risk factors in North Karelia with those in . . . [Kuopio] showed a highly significant net reduction in North Karelia of . . . 17 percent for men and 12 percent for women. Thus the programme was probably effective at least with regard to these intermediate objectives" (p. 1182). However, given the long exposure period for the development of CVD, a complete evaluation of the project will be dependent upon continuing follow-up of morbidity and mortality to assess future trends.

Commenting on the project, Wagner (1982) notes that while there were real reductions in risk factors in North Karelia, the background of generally declining CVD rates in industrialized countries poses the problem of identifying specific program effects, and of deciding "which CVD risk factors justify intervention and are likely to impact favorably on morbidity and mortality" (p. 52). In conceptualizing CVD as a community rather than individual problem however, the project has shown the feasibility of a community-oriented approach. Wagner also notes that, unlike most CVD prevention trials, the North Karelia project was politically and economically integrated into existing medical and public health resources.

Klos and Rosenstock (1982) comment that while the project attempted to combine many intervention strategies, the relative impact of the various components and their cost-effectiveness are difficult, if not impossible to evaluate. But as Wagner noted, the project was not designed as a field experiment but as a community attempt to reduce CVD rates, and was successful as such.

We have dwelt upon the North Karelia Project, even though it was a community- rather than a workplace-oriented intervention, because it illustrates the potential impact of a comprehensive approach to reduction of risk factors, incorporating technical, social, and cultural strategies. In so far as work organizations are in themselves communities, we have much to learn from the North Karelia and Stanford projects with regard to the principles underlying an effective approach to risk reduction. Most workplace interventions referred to in this book tend not to incorporate the elements that are found in the Stanford and Karelia projects. Essentially, such approaches seek to work within the dynamics of a culture—whether it is in a community or a company or both—in order to modify that culture. The "Live for Life" intervention at Johnson and Johnson (see chapter 9) is a good example of how these principles can be applied to a work organization through a social marketing approach (Wilbur and Garner 1984).

In 1972–1973 the U. S. National Heart and Lung Institute awarded funds to twenty-two clinical centers, laboratories, and a coordinating center for the Multiple Risk Factor Intervention Trial (MRFIT). As outlined by Sherwin and associates (1981), unlike the North Karelia Project, MRFIT was designed to evaluate the hypothesis that death from all causes and from CHD and fatal and non-fatal myocardial infarction (MI) could be reduced by lowering risk factors (smoking, serum cholesterol levels, and high blood pressure). The trial was to be carried out among men 35 to 57 years of age with no symptoms of CHD.

From 1974 to 1976, 361,662 employed men were voluntarily screened, leading to a sample of 12,866 participants considered to be at high risk for CHD on the basis of serum cholesterol, blood pressure, and smoking habits, and who agreed to conditions of MRFIT participation (for example, attempting to quit smoking). Participants were randomly assigned to either the Special Intervention (SI) program, or the Usual Care (UC) (i.e., under the care of his own physician) program. Participants in both programs were given annual examinations for the duration of the study, measuring risk factor status and clinical end points, and all received messages on nutrition and smoking as part of the orientation process.

SI participants could choose individual counseling (very few so choosing) or a group mode of intervention. Group techniques were used on the assumption of efficiency and effectiveness as forums for discussion of goals, fostering of group cohesiveness, and building of self-esteem. In the group mode the social support of participants' wives ("or homemakers") was seen as essential, and a precondition of group participation was wives' attendance at orientation and familiarity with materials (Benfari 1981, 439).

Although simultaneous reduction of all evident risks was the goal, SI participants were allowed to choose one of three SI programs: the Nutrition Intervention Program (a dietary calorie-reduction program), the Hypertension Intervention Program (using diet and drugs, monitoring compliance and side-effects), and the Smoking Intervention Program (behavioral techniques, including focal groups, systematic desensitization, contingency contracting, etc.). At each clinical center SI programs were supervised by either the chief nutritionist, smoking specialist, intervention physician, or hypertension coordinator, and a number of other health and social science professionals were available as resources. MRFIT staff members were assigned to individual participants as health counselors.

After the orientation session and randomization into SI or UC programs, those in groups attended nine more 1½- to 2-hour sessions, which included group discussions, various educational materials, films, and so on. After the tenth session (usually about four months after orientation), a regular four-month follow-up visit and a case conference, the Maintenance Program and Extended Intervention Program were made available to SI participants. The Maintenance Program was designed to reinforce success at reducing risk factors, and to minimize recidivism. The Extended Intervention Program was aimed at the group who had not yet succeeded in risk reduction, or who were recidivists. Individuals "could be in the Maintenance Program for one modality and in Extended Intervention for another. A totally successful participant was in the Maintenance Program for all modalities germane to his risk factor profile" (p. 440). The Extended Intervention and Maintenance Programs were in effect from 1974 until 1981.

In terms of meeting the goals of CVD mortality rate and risk reduction attributable to MRFIT intervention the results were somewhat disappointing.

The MRFIT Research Group (1982) reported that of the total SI and UC participants, 19 percent reported being prescribed antihypertensive medication at time of randomization into SI and UC groups. Six years later 58 percent of SI and 47 percent of UC men reported these prescriptions. The difference in UC and SI blood pressure at annual visits averaged 4 percent; after six years SI blood pressure was reduced by 12 percent, UC by 8 percent.

Initially, 59 percent of men were current smokers. After six years 50 percent of SI men had quit (46 percent thiocyanate-adjusted), and 29 percent of UC men had quit (29 percent thiocyanate-adjusted). The SI–UC differences in thiocyanate-adjusted quit rates at the twenty-two MRFIT centers ranged from 5 to 24 percent.

SI men showed lower reductions and UC men showed greater reductions in cholesterol levels than anticipated after six years with the UC–SI difference finally being only 50 percent of the goal.

The combined risk-factor reduction was reported as being "nominally at goal" at six years. It was expected that initial large differences would taper off with poor adherence. However, "the data show long-term maintenance of more modest initial differences" (p. 1470).

By the end of follow-up there had been a total of 260 UC deaths, substantially fewer than the expected 442. CHD mortality was lower in the SI group by a difference of 7.1 percent (statistically not significant). The death rate for all causes was 2.1 percent higher for SI men.

The subgroup analysis yielded significant results when including the 28 percent with abnormal baseline electrocardiograms (ECGs). For the group of hypertensive men with ECG abnormalities, there were 15 (65 percent) more SI than UC CHD deaths. The research group explained this as due to the "possibility that [these] participants . . . responded adversely to MRFIT intervention" (p. 1473). Excluding this 28 percent with abnormal ECG, there were fewer SI than UC deaths in all subgroups, and the UC group had a higher risk of mortality, but not significantly.

The investigators offer three explanations for their findings:

1. the program does not affect mortality from CHD;
2. the intervention does affect CHD mortality but the limited six-year term of the trial, and concurrent changes in the control group prevented its demonstration; and
3. efforts to lower the risk factors of blood cholesterol levels and smoking "may have reduced CHD mortality within subgroups of the SI cohort,

with a possibly unfavorable response to antihypertensive drug therapy in certain but not all hypertensive subjects" (p. 1465).

The last explanation was accepted by the investigators as the most plausible.

Oliver (1982) notes, however, that these subgroups constitute less than one-tenth of the total study population and "it seems unlikely that the small numbers with such an adverse response . . . could have diluted any important positive overall effect" (pp. 1065–66). Citing the parallel findings with unexpected favorable changes in the control group of the North Karelia Project, Oliver favors the second explanation. The MRFIT investigators acknowledge contributing elements such as the possible psychological motivation that UC men may have developed through participation in a project limited to high-risk individuals, the prevention measures that their own doctors may have initiated, or the health education and sensitization impact of participation requirements (annual visits to clinics for monitoring of risks). It is also possible that, because many of the participants in UC and SI groups came from the same workplace, discussion and competition surrounding the trial may have developed.

Oliver also suggests that not only is the incubation period for CHD longer than the six-year trial duration, but the intervention may have no effect on CHD. He cites our lack of knowledge of regression of coronary lesions regardless of risk-factor control and suggests that the usual risk factors may not be those that are most relevant for people who have moderate or advanced obstructive arterial disease.

From the World Health Organization European trial in industry, Oliver states that the "interim report on the changes in risk factors after two and four years of intervention makes depressing reading. Changes were smaller than expected and not completely consistent or sustained" (p. 1066). However, CHD incidence in an Oslo study (cited by Oliver) showed a reduction with cessation of smoking and a dietary intervention in nonhypertensive high-risk men.

Oliver has raised an important question for this area of research, namely: "Does control of risk factors prevent coronary heart disease?" (p. 1065). At this point, the evidence does not point unequivocally to reductions in CHD resulting from risk-factor control. While a "preventive life style" may be suggested by common sense, there have been no studies with absolute proof that changing one's life style ensures health or long life. This is not to imply that demonstration of such a relationship is impossible—it may simply be too early to tell. On the other hand, while these studies have provided valuable information with regard to effective health education, the changes that they have initiated have been primarily behavioral. These studies (except for the North Karelia Project) have also focused on human risk factors, to the exclusion of technological, occupational, and environmental risks, which might also be included as components of "life style."

With regard to the control of hypertension (as a risk factor for CHD and CVD) through drug therapy, Wing (1984) has raised a contentious issue that underlies the medical approach to a variety of diseases and disabilities, the incidence of which can be traced at least in some measure to socioeconomic conditions. Wing argues that the availability of a partial medical solution (control of hypertension through drug therapy) should not be allowed to divert attention away from the environmental conditions that raise the risk of developing high blood pressure. Thus, education, income, occupation, race, and region are all factors that influence the chances of developing hypertension through their impact upon nutrition, exercise patterns, stress levels, smoking and drinking, work patterns, and so forth. In a real sense, then, socioeconomic conditions are a class of risk factors in themselves.

Socioeconomic and Sociobehavioral Factors in the Etiology of Disease Outcomes

In their overview of the literature on the social etiology of chronic disease, McQueen and Siegrist (1982) summarize current epidemiological research on CVD, cancer, and multiple disease outcomes. In all three of these outcome categories psycho-social stress was found to be an important concept. However, the factors postulated as contributory to such stress cover a very wide range of socioeconomic and sociobehavioral variables. By far, the largest body of research is in the field of coronary heart disease.

McQueen and Siegrist note that economic change is seen as a significant factor in the "health status of occupational groups and their families with special impact on CHD" (p. 355). The components of socioeconomic status (SES) (measured by variables of income, occupation, and education), and changes in SES are each factors in the etiology of CHD, each generating a specific stress. McQueen and Siegrist cite evidence from several studies showing that periods of economic growth foster stress due to overwork, while economic recession brings insecurity, weakening of socio-emotional support, increased marital discord and vulnerability (Brenner et al. 1980; cited in McQueen and Siegrist 1982, 355). Other researchers have concluded that increased stress and "weaker coping capabilities" are linked to higher CHD incidence in lower SES groups. These "weaker coping capabilities," of course, may not refer so much to psycho-emotional "inner" resources, as to knowledge of, and access to those professional and financial sources of help taken for granted by the middle class. This lack of knowledge may be stressful in itself. Lower occupational groups have also shown increased mortality from CHD, and lower educational groups show three times the risk of sudden cardiac death after first myocardial infarction. These groups are also most likely to be affected by stressors such as noise and chemicals that have been linked directly with CHD.

McQueen and Siegrist note that most of the research on work stress and CHD focuses on the individual's experience of work stress, mental workload stressors in particular. Although this body of research lacks a consistent theoretical approach, and in parts suffers from vague measures of work stress, significant relationships have been found between it and CHD incidence (Theorell et al. 1975, cited in McQueen and Siegrist 1982, 356). For example, Kornitzer's 1979 job stress scale scores showed a highly significant relationship to the presence of angina pectoris in a study of 2,000 industrial workers (Kornitzer et al. 1981, cited in McQueen and Siegrist 1982, 356).

Closely related to the body of work just discussed is the growing body of research in coronary prone behavior (CPB). This behavior is sometimes attributed to what has become known as the "Type A" personality. It is characterized by an almost frenzied competitiveness, an inability to relax, driving ambition, a sense of time urgency, impatience, and an insistence upon practicality over philosophy. CPB is more prevalent among white-collar workers, with several stress-related explanations offered:

1. CPB is related to professional competition—the struggle for upward mobility and a self-selection process based on "survival of the fittest" whereby only Type A personalities attain middle-echelon positions. However, McQueen and Siegrist cite Kasl's precaution that this explanation comes from a sex and occupationally biased sample. The all-male sample was primarily middle- and upper-level managerial-scientific workers, with only 10 percent being blue-collar (Kasl 1978, cited in McQueen and Siegrist 1982, 356). Hence evidence from a sample that is 90 percent middle- and upper-echelon cannot conclusively be explained by the "professional self-selection" possibility.

2. CPB subjects may automatically perceive their workloads as stressful, but even so it seems likely that such people are more disposed to take on heavier responsibilities and more demanding schedules.

3. CPB may reflect coping strategies elicited by the work setting. However, while experimental studies lend support to this explanation, few studies in "real life" settings do so. The most critical elements in eliciting CPB appear to be instability of work autonomy and control, interruptions and time pressures on the job, and inconsistent demands. While this area of research has focused on male workers, these job characteristics bear a strong resemblance to those of clerical work—the group of workers in the Framingham data shown to have the highest rates of CHD (Haynes and Feinleib 1980).

In an analysis of the seven causes of death for which male mortality in the United States exceeded female mortality by 100 percent, Waldron (1976) points out that while some of the increased sex differential in mortality in the twentieth century (from a female advantage of two years life expectancy in 1920 to eight years in 1970) is due to the decline in maternal mortality and uterine cancer, it has also been influenced by cultural factors, such as smoking, driving, and taking care of one's health. For example, among never-smokers, CHD mortality rates for men exceed those for women by 350 percent; among a sample including smokers, male CHD mortality rates are excessive by 650 percent.

Waldron points to evidence that suggests that men who display CPB are more likely to develop or die from CHD, while women who already have CHD are more likely to display CPB than are controls. Citing evidence that women who display aggressive, competitive behavior were more likely to develop CHD than other women or than less aggressive, competitive men, she concludes that this type of behavior may be more important to CHD risk than sex per se. While Waldron sees the pressures of the workplace as contributing to the development of CPB she also sees this type of behavior as integral to the male role in our culture. Sex differences in competitiveness are developed in childhood when boys are pushed to occupational achievement and girls are encouraged to achieve success in the family. It seems that socio-economic and cultural pressures indirectly contribute to men's higher risk of CHD through reinforcement of CPB.

In terms of respiratory diseases, male mortality rates are six times higher than women's for lung cancer, and five times higher for emphysema—both disease outcomes that are related to cigarette smoking. Comparing death rates for emphysema among smokers and nonsmokers, rates are elevated sevenfold among male smokers, but only fivefold among women smokers. Similarly, lung cancer death rates are elevated ninefold among men, but only twofold among women. (This is partly because women smokers smoke fewer cigarettes, smoke less of each cigarette, and inhale less.) Another major contributing factor in male respiratory disease rates, however, is their higher likelihood of exposure to industrial carcinogens such as asbestos and metallic dust and fumes. Waldron cites evidence that roughly one out of ten male lung cancer deaths can be linked to suspected and established industrial carcinogens (p. 354).

While lung cancer is more common in men than women, women are more susceptible to breast cancer and cancers of the reproductive organs. Men's higher cancer mortality, however, is not seen as due to genetic differences or sex hormones.

Waldron reviews other causes of death for which the male mortality rate is excessive. Death rates for motor vehicle accidents, for example, are about three times higher for men than for women. Men are involved in 30 percent more accidents per mile driven, and some evidence shows that while men often drive at rush hours, they also have less safe driving habits in general. Roughly half of male excess mortality is due to work-related accidents, again a reflection of the male societal role in dangerous or unhealthy occupations. Drowning accounts for one-third of the excess, and accidents involving firearms are five times more common among men. Waldron attributes these figures to behaviors that are integral to the male role: doing sometimes hazardous work, driving, using guns, "being adventurous and acting unafraid" (p. 354). Further, many accidents involve alcohol consumption, which is more common among teenage boys and men than women.

While women are twice as likely to unsuccessfully attempt suicide, men actually commit suicide three times more often than women. In years of economic recession, male suicide rates increase by 9.5 percent per year, while women's suicide rates increase by only 2.9 percent per year (p. 355). This may be partially due to the stress of competition for jobs experienced by men, coupled with the fact that men are less likely to seek help during periods of suicidal contemplation.

In general, mortality rates are higher for males but morbidity rates are higher for females. "Women report more symptoms, visit doctors more often and more often restrict their usual activities or spend a day in bed because of illness" (p. 356). Women also take better care of their health in that they use preventive services more frequently, although they do delay seeking medical attention after first symptoms of cancer and myocardial infarction for as long as men do.

Waldron concludes that reductions in male mortality might be achieved by cultural and behavioral changes. The development of CPB, for example, should be addressed at earlier ages, before damage is done to coronary arteries in middle age. She also takes a structural rather than psychological approach to the causes of and solutions to CPB. An important social structural factor in the adoption of CPB is the "scarcity of satisfying jobs and the large differentials in pay and intrinsic rewards . . . (of available) jobs" (p. 358). She suggests a restructuring of organizations with responsibility and profit sharing to alleviate these inherent pressures (see chapter 8 for further discussion).

While CHD is not usually thought of as a "male problem" in the way that uterine cancer, for example, is a "female problem," the fact remains that men are more susceptible to, and more often die from heart disease than do women. Men are also found to display CPB more often, and are the focus of CHD/CPB research (MRFIT, for example, spent millions of dollars studying risk factors in men only).

Tolson (1977) has approached the same behavior ("latent hostility, work addiction, time urgency, . . . resistance to relaxation, speed and impatience, . . . a desire to dominate" [McQueen and Siegrist 1982, 356]) in a group of working- and middle-class British males from a different perspective and arrived at a more sociological reading of both the behavior and the culture that fosters it.

Tolson would argue that CPB is the symptom of a crisis in the experience of masculinity. He argues that "in Western, industrialized, capitalist societies, definitions of masculinity are bound up with definitions of work" (p. 12), with the qualities of a successful worker being those of a successful man. In addition, the traditional definition of masculinity involves a patriarchal role in the family setting.

Both traditional sources of masculine identity have taken a considerable battering in the last twenty-five years. The threat of unemployment and the probability that few will attain the high-status positions to which they aspire coupled with rapidly changing expectations about the role of men in the family create a situation in which males who cannot or will not change their behavior or their expectations feel increasingly trapped. This perception may be influential in the aggravation of CPB. The process that links the feeling of being trapped with increased manifestation of CPB appears to involve a kind of panic-stricken response to being blocked or frustrated.

In the light of the foregoing, it is not surprising that research has yielded no significant evidence of consistent CPB personality types. Whether the foundations of CPB are in social structure (as Tolson and Waldron would suggest) or in personality composition, they are not easily defined and influenced. Interventions that have attempted to change behavior (MRFIT and the North Karelia Project) have focused on diet and exercise; while the Stanford heart study attempted to change attitudes it was focused on cardiovascular health per se, not on unspoken motivations to unhealthy behavior. It is also very interesting that one of the reasons given for the failure of MRFIT to show significant results is the competition that developed at the workplace among the men involved in the trial. If the intervention is viewed as a competition it may be stressful and unhealthy in itself.

McQueen and Siegrist are critical of the social epidemiological research that has been done, finding definitions and descriptions of disease frequently too vague, measurement of psycho-social variables too imprecise and varied, data collection and statistical analysis too simplistic, and longitudinal studies too rare. The complexity of social variables, which are defined and analysed as if they were biological variables, is often ignored.

A second weakness of the traditional epidemiological approach, McQueen and Siegrist suggest, lies in the question of causality. Correlation bivariate tables are often presented when the use of factor analysis and path analysis in examining multiple variables would add further depth to the approach.

Moving to the etiology of cancer, McQueen and Siegrist point out that there is considerable ignorance as to the role of psycho-social factors (i.e., whether they influence only certain types of cancer or have an impact only on certain stages of its development). Given increasing interest in the area, they suggest that more research be done on the relationship of cancer to "no-exit" situations of helplessness, long-term emotional repression in the family, and traumatic childhood experiences.

With regard to the genesis of cancer, Cunningham (1985) has reviewed the literature regarding the relationship between this disease and psycho-emotional factors. This area of research has not yielded definitive conclusions. Of particular interest, however, is the link between cancer and depressive affect, repression of emotion, submissiveness, and self-derogation. These factors are believed to interact in a complex manner with social support. The putative pathway between the influence of all these variables and cancer is hypothesized to be the immune system of the human body. There is some reason to believe that this system is disrupted or defeated by certain types of chronically recurring emotional events. The view that chronic stress may catalyze such a course of events is not without foundation.

Consistent with the kinds of criticism leveled by McQueen and Siegrist, McQueen and Celentano (1982) point out that most of the literature addressing social and biological factors in chronic disease causality focuses on single disease entities. In an effort to explore the complexity of the etiological process, they take as an example the relationship between high blood pressure and excessive alcohol use. To date, the relationship between the two outcomes has primarily been observed at "end-points," with the *process* of interaction between the two variables neither observed nor understood. In general alcohol use is thought to occur prior to and to act as a cause of hypertension. McQueen and Celentano argue however that the two outcomes may occur simultaneously, interacting with stress and social support or the absence of it (p. 405). They emphasize the need to view etiology as a process and to study the variables as continuous.

Work-Related Stress

An argument can be made that any unsafe or hazardous working conditions are inherently stressful and should be a concern for public health professionals and health educators (Mansfield 1982). Support for this argument, however, would necessitate a review of the vast literature on unsafe work. Nonetheless, we acknowledge that certain occupations may be inherently stressful due to unhealthy or hazardous working conditions. Mining, heavy industrial work, construction work, monotonous and tedious assembly-line work have long been accepted as dangerous (Fitch 1982; Bell 1970; War and Wall 1975), while more recent technological change has added new occupations to the list: telephone operators, office workers who must work in "white noise" and under fluorescent lights, VDT operators, and synthetic chemical workers, among others (Epstein 1979).

While in the past women were legislatively "protected" from the hazards of (high-paying) heavy industrial work, technological innovations in clerical ("women's") work, and women's reentry into hazardous, traditional male industrial work have raised new cause for concern about women's health. The potential health hazards of these occupations have included a variety of disease outcomes (heart disease and cancer, for example), chronic conditions and disabilities (including hearing loss and eye conditions), and reproductive problems (such as miscarriage and birth defects—dangers that sometimes lead employers to suggest that women workers must choose between sterilization and losing their jobs) (see Mansfield 1982).

In her comprehensive review of the literature on women and work-related stress, Haw (1982) made several general observations:

1. In the ten years prior to her review, studies on men's work-related stress outnumbered those on women's by six to one.

2. Because studies on women's work-related stress tend not to focus on specific work-environment stressors, as do those on men's, the implication is that it is paid work per se that is stressful for women.

3. Women's roles at work and in the home are addressed in studies on women and work stress, whereas studies on men seldom address home and family roles. The suggestion is that role demands vary for workers, depending on their sex.

4. Related to the above, over 33 percent of studies that Haw reviewed detail the expansion of role demands placed on working women. While many indicate women's lower job satisfaction and more negative feelings about work compared to men, few relate these factors to disease outcomes (such as CVD) or psycho-emotional illnesses (p. 135).

Haw's observations raise several points. First, with women's growing labor force participation, their advantage over men in mortality rates may decline. While the proportion of studies focusing on male and female work-related stress may have once reflected differences in the types of situations faced by men and women in the paid labor force, one study of women for every six of men is no longer adequate. Women's paid work participation rates in the United States have increased from 33 percent of women over 16 in 1950 to 53 percent in 1975. In 1969 only 29 percent of those women had children under the age of 6 as compared to 37 percent in 1975 and 54 percent in 1982 (Haw, p. 132; Mansfield 1982, 6). Trends in Canada and Britain are in much the same direction (Armstrong and Armstrong 1978; Mackie and Patullo 1977).

Second, the implication that women may find employment per se to be stressful suggests several assumptions: (1) that women find work outside the home stressful in itself because it is somehow contrary to women's nature—

that women are mentally and/or emotionally and/or physically less capable of paid work than are men and therefore find it stressful; (2) that given the full-time responsibilities women carry in the home, the (second) full-time job is more stressful for women than it is for men; or (3) that the low-status, low skilled, low-paying jobs that most women hold (Mackie and Patullo 1977; Connelly 1978; Armstrong and Armstrong 1978; Mansfield 1982, among others) by definition involve higher levels of stress.

Third, as discussed earlier, it may be precisely the negotiation of public and private roles, neglected in the literature on men, that is a key component of work-related stress and outcomes such as CVD and CPB. The focus on conflicts of dual roles may be as valid for men as it is for women. Underlying much of the research on male work-related stress is an assumption of the validity of the myth that for men home life is a peaceful refuge away from the pressures of the public domain. While home may be private, for many people across classes it is also a source of extreme emotional conflict rooted in a range of possible problems such as financial pressures, ill health, housework and childcare conflicts, alcohol and drug abuse by family members, parent–child conflicts, and even violence. The problems men may face in the private sphere are seldom, if ever, even acknowledged in stress research, yet are assumed to exist and to be important for the women with whom they share their lives.

Finally, from Haw's observations, research that links multiple role demands to such outcomes as CVD cannot claim the external validity it often seems to assume when that research (such as MRFIT) focuses exclusively on men (p. 132).

The tendency for researchers investigating the links between stress and future morbidity to deal with the mediating variables discussed by Haw in an inconsistent manner, if at all, has led to a somewhat confusing body of data on the relationships. Haw's review of the literature linking type and conditions of employment to stress and morbidity yields contradictory results that are probably traceable to the different conceptualizations of the relationships in the various studies considered. Some writers, such as Roskies and Lazarus (1980) and Lumsden (1981) have called for a *transactional* approach to the study of stress in which the focus of investigation is upon the coping processes that are used by individuals and groups to deal with stressors. The phenomenological meaning of stress is at the center of this approach, necessitating a method of study in which the individual realities of workers are given form and expression. Currently, the correlations that exist between stress (usually defined according to the presence of external stressors often conceptualized as "life events") and future morbidity are frequently positive and of statistical significance. However, the correlations are not sufficiently robust that they can be used to *predict* disease outcomes from knowledge of stressor exposure alone. While it seems clear that relationships exist between stressors and disease, they remain to be specified in such a way that effective interventions can be developed to interrupt the process (see chapters 6 and 8).

House (1974) noted that the task of understanding these relationships involves a study of the personal, cultural, and social factors influencing the extent to which people interpret information about external events as threatening or otherwise.

The Philosophical and Political Debate Surrounding Intervention

If we accept the premise that relationships exist between current life style and future health, we are then confronted with a major debate over the appropriate locus of preventive and remedial action. The debate hinges upon two perspectives that combine elements of philosophy and politics. These are, at their extremes:

1. That individuals are responsible for their health, that individual ill-health is an unfair social and economic burden on society, that most illness is "life-style–related," and therefore the individual is the target at which health promotion/education should be aimed.

2. Opposingly, that holding individuals responsible for their own health is a sophisticated form of victim blaming that diverts attention from the sources of disease and illness in the physical and social environment, that the focus on individual responsibility creates social passivity and, paradoxically a further dependence on an inefficient health care system.

For the sake of convenience we will here label the two perspectives, respectively, as the individual model and environmental model.

The courses of action stemming from each approach vary accordingly. Interventions within the individual model are designed to reduce accepted risk factors such as smoking and high blood pressure. Approaches within the environmental model are aimed at placing the individual's awareness of his or her health within a cultural and socioeconomic context that sees traditional risk factors as responses to more deeply rooted structural problems.

In a sense, the chronology of the intervention debate has paralleled the chronology of the research on the relationship itself. That is, the philosophical debate develops alongside practical, concrete actions in the same way that intervention trials coincide with theory and research.

In North America the individual model seems to have preceded and, not surprisingly, gained more popularity than the environmental model. The latter appears to have developed as a response to the former. It should be noted that health education professionals as implementors of some of these programs, play a key role in articulating these approaches to the public, in putting the findings of the research discussed earlier in this chapter (the work of HPL, MRFIT, etc.) into concrete practice and may be responsible for far-reaching consequences in terms of life-style change or demands for health reforms. It will be useful, therefore, to examine the historical roots of the Individual Model in light of how health education has developed in order to understand more fully what the environmental approach is responding to.

In their review of the "historical roots, ideological perspectives and structural constraints" of the health education profession Brown and Margo (1978) cite the school health movement (c. 1925–35) as the beginnings of health education. Concurrent physician control of public health boards ensured the medical model as the paradigm through which to see disease and "maladaptive" social problems. By the 1940s the health educator's role as teacher was not only to reduce the spread of disease through public education, but also to direct the public to "'appropriate' use of medical services" (p. 4).

Brown and Margo suggest that rather than helping people to change those social and personal environments that foster attitudes "perceived by health professionals to be careless, irrational or both," the goal of health education has become *behavior change* (p. 9). This, of course, is a critical perspective on the historical background to the present individual intervention approach. However, Brown and Margo's point that use of the health care system was encouraged by health educators/promoters is central to the problem now under debate.

Today, as we have noted earlier with reference to the work of Zook and Moore (1980), it seems that there are those who are "not making enough use of services [while] others are accused . . . of 'over-utilizing' them" (Brown and Margo, p. 5). Those who "over-utilize" services are seen as imposing an unfair cost burden on the system, and on those healthy individuals who have little need of professional health care. "Poor health habits," "adverse life style," and "persistent refusal to follow physician's advice" are emphasized by Zook and Moore as characteristics of high-cost users of the health care system.

At a time when, as Roemer (1984) points out, national expenditures on health are steadily rising and lead to efforts at "improved efficiency" through health restraints, cost-efficiency of the medical care system vs. potential benefits of healthy individual life styles emerges as a recurring theme in the literature and may in fact be the crux of the debate. For example, several statements issued by governments have expressed concerns about health care and its attendant costs. In the United States, *The Forward Plan for Health, FY 1978–82* stated:

In the absence of a major scientific breakthrough, (e.g., a cancer cure), further expansion of the Nation's health system is likely to produce only marginal increases in the overall health status of the American people. (U.S. Dept. Health, Education, and Welfare 1976, cited in Shapiro, 1977, 291).

In his foreword to *Healthy People: The Surgeon General's Report on Disease Prevention and Health Promotion,* Joseph Califano writes that

> ... we are a long, long way from ... national commitment to ... good personal health habits ... in America. And meanwhile, indulgence in 'private' excesses has results that are far from private. Public expenditures for health care that now consume eleven cents of every tax dollar are only one of those results. (U. S. Dept. of Health, Education, and Welfare 1979, p. *ix*)

As we noted earlier former U. S. president Jimmy Carter expressed concern that in 1977 "the average American worker is now devoting one month's worth of his yearly salary just to pay for medical care costs" (cited in Crawford 1977, 665), and medical expenditures have claimed an increasing proportion of the U. S. GNP (from 5.3 percent in 1960 to 9.4 percent in 1980, as reported by Parkinson et al. 1982, 1). Sidel (1979) reports that in the United States medical care costs see "fifty cents out of every dollar going to hospital and nursing home care and less than five cents of each dollar spent on health protection or promotion" (p. 235).

Many reports and statements, including Marc LaLonde's *A New Perspective on the Health of Canadians* (1974), acknowledge that illness and disease can be caused or influenced by environmental problems but choose to focus on the possibilities for improved health through personal life style. Danaher (1980), for example, first cites statistical estimates of the number of smoking-related deaths due to various causes (e.g., "220,000 deaths from heart disease"), then estimates the statistical accountability of smoking for various diseases (e.g., "40 percent of all respiratory disease"), and provides estimates of costs that can be attributed to smoking (e.g., "$3 per day per smoking employee"). He then notes the growth of smoking cessation programs in the workplace attributable to a recognition of these costs on the part of business and industry (see also chapter 4).

Alderman, Greene, and Flynn (1982), discussing hypertension control programs, state that in 1976 "cardiovascular diseases consumed, in direct and indirect costs, some $50 billion, or 20 percent of all health-related expenditures" (p. 162). Most hypertension-control programs, of course, assume that individual action, with medical help, can bring about significant improvement as indeed it can (see Foote and Erfurt 1983, 1984; Erfurt and Foote 1984).

Schwartz (1982) outlines stress management programs in occupational settings and provides a brief background summary of sources of occupational stress, with some evidence of cost beneficial program results. Manuso (cited by Schwartz) estimates that individual and group-oriented stress management techniques average an annual corporate savings of $2,861.82 per employee with chronic anxiety or headache (p. 244).

Reports on the development of these intervention programs generally tend to cite briefly evidence in support of the relationship between life style, or "health habits" or "health practices," such as smoking, alcohol use, and so on, and future health. Little mention is made of any problems in research on the social epidemiological basis of the relationship. While few would argue that the aforementioned health hazards (smoking, hypertension, and stress, among others not mentioned here) are invalid, our point is that this genre of research is based on an accepted paradigm of the life style–future health relationship, the structural dynamics of which have yet to be established.

One of the tools sometimes used in individual intervention programs is the *health hazard appraisal*, or the *health risk appraisal* (HHA/HRA). HHA/HRA is an instrument used by health educators to stimulate life style change through an appraisal of individuals' health risks and is a product of the integration of health education, medicine, and technology. Usually, a questionnaire is used to assess the individual's health practices; then his or her profile is compared to actuarial tables outlining five-year mortality rates according to age and sex. The client is assigned a "risk age" and an "achievable age" based on the information provided. For example, a 55-year-old man with poor health habits and a family history of heart disease might be assigned a risk age of 60. This means that the individual has the health characteristics of a man of 60, not that he will probably die at that age. He will then be advised on healthy life-style changes that could help him in his goal toward the achievable age of 50.

Several criticisms have been made of HHA/HRA. Following others, Goetz, Duff, and Bernstein (1980) suggest that data bases for mortality rates must be updated every ten years, and must also account for geographic variation in mortality rates. More problematic in estimating risk factors is the data base for morbidity incidence. Apart from the Framingham CHD incidence data "only fragmented data exist, approximating to varying degrees the data required for risk estimation" (p. 123).

Wagner and associates (1982), on the other hand, disagree with even the use of Framingham data on a widespread basis of risk estimation. They point out that it involved largely middle-aged middle-class whites, "yet [the] findings are being used to predict the risk of Blacks, Hispanics, Native Americans, teenagers, and other dissimilar groups. In a few cases, the extrapolations produce nonsensical risk values" (p. 349). Wagner and associates also point out that some of those characteristics and relationships addressed by HHA/HRA are scientifically controversial—the "predictive importance" of physical activity in CHD death, for example. Also cited as problematic are the evidence that cholesterol reductions lower risk of CHD death, and the usefulness of including characteristics such as family history that cannot be altered.

Secondly, Wagner and associates point out problems in applying HHA/HRA to younger clients given the slow increase in ten-year death risks for those under thirty-five, difficulties in using risk estimates based on studies of older populations, and the "lack of salience to young people of their long-term mortality prospects" (p. 349). Finally, Wagner and associates note that some risk factors for death due to causes such as suicide and homicide are not well or easily specified and quantified.

While only a minority of intervention programs employ the HHA/HRA, its use is growing and it is now available in microcomputer programs. Clients interact with the computer and results are produced immediately. It should be pointed out, however, that recent evidence suggests that the use of computers in medical consultation can actually increase patient stress, especially in those who are already highly stressed (Cruickshank 1982). HHA/HRA not only concretizes the Individual Model's focus on individual responsibility for health through behavior change, but appears to be a health promotion tool that is keeping pace with technology; in reality, however, it is based on contradictory evidence and incomplete research.

The basic criticisms that have been directed at the Individual Model of health promotion may be seen as devolving into the following points:

1. The individual intervention approach holds individuals responsible for their health while downplaying environmental sources of disease, thereby blaming individuals for sources of illness that are presently beyond their control.

2. Following this, the approach encourages a myopic self-centered view of disease and health, deflecting attention away from the possibility of collective political and social action in health-related issues.

3. The approach is strongly middle-class–biased with little attention paid to the concerns and realities of those who are not.

4. The approach focuses exclusively on the individual, as opposed to any social special interest groups.

Crawford (1980), a major spokesperson of the environmental perspective, has developed a critique of individual intervention as a victim-blaming ideology that he labels "healthism." He includes self-help and self-care movements in the trend toward medicalization, wherein "health" emerges as the principal solution to personal and social problems. As "health" becomes the "pan-value" of our society, Crawford observes a new moralism in which personal health becomes the standard against which to measure the worth of any action or venture. Even emotions and attitudes can become medical risks and a new role, "the potential sick role," develops wherein one assumes a moral obligation to avoid sickness through living the right kind of life.

Crawford's argument is supported by Tesh (1984) who offers an analysis of how stress research can be used politically to suit the interests of opposing groups, in this case the U. S. Professional Air Traffic Controllers Organization (PATCO) and the Federal Aviation Administration (FAA). Tesh notes how, in general, stress theory has enabled occupational safety and health committees to include the social and interpersonal occupational environment, including democratic working conditions, in their range of concerns and demands. However, in the case of the 1981 PATCO strike, the focus on stress research and lack of agreement within the scientific community on the definition of "stress" created practical problems in achieving the new goals (p. 571).

PATCO spokespeople defined stress as those job conditions and "tensions inherent in air traffic control," and FAA supporters defined stress in terms of its biological outcomes, which may or may not result from the stressors of work in air traffic control (p. 576). PATCO argued that if controllers were expected to endure the stress of such working conditions, then they should be adequately compensated for it. The FAA argued that monetary compensation would not solve the problem created by employing controllers who as individuals were psychologically and physiologically unable to cope with the demands of the job, that is, by employing controllers who were stressed. The striking controllers were fired by the Reagan administration, their union was dissolved, and new controllers were hired to replace PATCO members.

As Tesh points out the FAA's separation of worker from working conditions denies the holistic approach of interdependence of mind and body upon which Selye's ground-breaking work on stress was based. The emphasis on individual controller's psycho-physiological indicators of stress points to precisely the victim-blaming uses of this field of research of which Crawford warns.

Katz and Levin (1980) summarize Crawford's position as a "powerful warning against the use of 'victim-blaming' ideology as a political and social weapon to still demands for basic structural change in the health care system" (p. 329). They acknowledge the validity of Crawford's class analysis of the roots of medical ideology and its use by powerful interests in various fields "to maintain a status quo of minimalist services and maximalist personal privilege" (p. 329).

Katz and Levin point out, however, that self-help health groups are in fact wholly based "on the idea of access to medical care as a *right* and therefore contradict any movement towards the erosion of health care rights" (p. 334, italics in original). In their opinion the self-help/self-care movements are a grass-roots mass response of health care consumers to the "medicalization" outlined by Crawford and to health interventions framed within the Individual Model. It is important to be aware that in so far as self-help is directed toward obtaining better medical care it is a politically active movement. However, not all self-help is politically oriented; the objectives of some groups are geared more toward freedom from medical control. In that sense, while it makes a political statement, it is doing so while stepping outside the boundaries of the medical realm.

Some writers have argued that otherwise conventional health education and interventions can be developed in such a way as to incorporate political awareness and thus counteract the tendency of programs within the Individual Model to assume a victim-blaming posture. Allison (1982), for example, points to attempts at health education in San Francisco and Toronto that incorporated Freire's principles of "critical consciousness" (Freire 1971; 1973) in such a way as to promote an awareness of the structural origins of poor health and health practices. Similarly, such programs can explore ways in which action can be taken to modify some of these structural problems.

Allison agrees with Minkler and Cox (1980) that the lack of a "sense of community" in North American society leaves Freire's philosophy "difficult to implement." They suggest however that the geographically based sense of cohesion in other cultures is often replaced here by "'functional' common interest communities, such as the women's movement" (p. 13).

Indeed, with reference to the women's movement, evidence presented by Simmons, Kay, and Regan (1984) supports Allison. They found that some U. S. women's health groups (32 of 123 possible groups in their sample) indicated a politically conscious alternative to the traditional "hierarchical" organization of health care and appeared likely to foster self-responsibility, rather than social control. They also found a community-based continuity in these organizations with their length of existence ranging from 3 to 12 years, the average being 8.1 years (p. 629).

It may be useful for readers to juxtapose the foregoing comments with observations made in chapters 5 and 6 where two experimental health promotion programs are evaluated. In both cases, the issue of empowerment is raised with regard to individual control over personal health and other major areas of life. In both cases, attempts were made to build into the programs the skills and cognitions required to engender a sense of competence or self-efficacy. The results in that regard were encouraging, indicating at the same time areas where improvements can and must be made. In light of the criticisms aimed at the individual model that we have just reviewed, it may seem a small thing and an inadequate response to devise programs, such as those described in chapters 5 and 6, that emphasize individual empowerment in quite narrowly defined areas of personal and familial health. Yet we would argue that this approach can provide a basic form of self-defense to participants through facilitating improvements in health-competence. When these improvements have both mental and physical health aspects, participants will feel physically more able and psychologically more ready to tackle larger more complex problems that bear upon their ability to make further changes. Having said this, we hasten to add that programs and interventions within the individual model that try to incorporate elements to encourage and facilitate empowerment are few and far between. In addition, the methods required to bring about such results are not developed within the health-allied professions. As will be apparent from the foregoing, there is likely to be considerable debate over the desirability and feasibility of incorporating empowerment components into health programs particularly in the workplace. Chapter 9 deals with some aspects of these issues as they relate to self-help and social support in organizations.

Conclusion: Implications for the Design of Health Promotion Programs in the Workplace

The evidence that we have reviewed in this chapter provides the basis of a rationale for the adoption of health promotion interventions in the workplace. This rationale assumes that employers are interested in keeping their employees in better health and that this objective may be achieved through helping workers to modify their life styles. However, a set of principles or guidelines is emerging that should influence the manner in which such interventions are planned and carried out. We might think of these guidelines as pertaining to both the effectiveness and ethics of health promotion. Thus, a growing body of research data suggests that health-related practices are significantly interdependent and in turn are heavily influenced by factors outside the immediate control of the individual. It has been shown that attempts to modify any one of these health-related practices without taking into account the others meet with limited success. Analysis of these results suggests that it will be necessary to approach the modification of specific risk factors in the context of more comprehensive approaches to the improvement of health and well-being. These approaches will deal with not only the interrelatedness of apparently specific health practices but will do so in the context of the prerequisite mental health conditions required to undertake significant changes. Such conditions include a sound ability to manage or cope with stress and a sense of being able to control at least some aspects of one's immediate environment as it relates to health. However, it will also be necessary to recognize the impact of external factors such as the work environment, the health orientation of organizations (or lack of it), and the macro-social influences on behavior that have health consequences, such as socioeconomic class and the manner in which gender identity is developed in our culture. An examination of these external factors should be built into programs and other levels of intervention aimed at modification or reinforcement of health practices. Related to this strategy, however, is the need for communities and workplace organizations to assume partial ownership of the responsibility to protect, or at least not abuse the well-being of their citizens and employees and to adopt policies and practices conducive to good physical and mental health.

These guidelines for health promotion are clearly going to be contentious. Indeed, if the arguments outlined in this chapter provoke discussion they will have served their purpose. To distill the case presented here, we are proposing an extension of the role of health promotion practitioners to include the following components:

1. The introduction of critical consciousness-raising components into program and course materials so that participants become aware of the constraints that society and the organizations by which they are employed place upon them with regard to personal choice about health-related behavior.

2. The introduction of methods for health improvement that take account of these constraints by emphasizing the need for collective action in some cases to buffer or reduce their impact. This could take the form of developing self-help/self-care or social support groups around specific health concerns such as nutrition, fitness, stress, heart problems, back problems, and so on. Alternatively, it could involve strategies that bear upon health indirectly—for example, the formation of child daycare support networks that alleviate the stresses of time pressure, schedule coordination, and worry about children's safety and well-being.

3. The provision of feedback to management and union about the impact of organizational stressors upon the mental and physical health of workers. This assumes a monitoring role with regard to these stressors.

It is clear that these role extensions can be brought about only in situations where senior management in some way endorses the principles underlying the approach. Without this endorsement, the activities of health promotion professionals become analogous to those of guerrillas, a situation that would not be tolerated for long. Even with the blessing of senior management, some rough encounters may be expected as health professionals provide feedback and urge change in areas pertaining to the organization of work or running the business. To ignore the impact of socioeconomic, cultural, and organizational influences upon individual health choices, however, is to court ineffectiveness in the operation of health promotion programs, particularly where the work force in question occupy predominantly middle- to low-level positions.

3

The Prevalence of Mental Health Problems and the Need for Employee Assistance and Health Promotion Programs: Case Studies

with L. Mark Poudrier

I n the two previous chapters, we reviewed the rationale for broad-based employee assistance programs and proposed that even the greater breadth afforded by these more recent approaches is inadequate to deal with the range of health-threatening conditions that are likely to arise among workers. It was further suggested that health promotion programs could play an important role in extending the range of responses to these conditions. In this chapter, we examine in greater detail some aspects of the mental health conditions that give rise to the need for broad-based EAPs and HPPs. The observations reported here were occasioned largely by a series of studies carried out by the authors acting on behalf of a provincial committee charged with the responsibility for implementing voluntary EAPs in selected government workplaces in Ontario. The full title of this implementing group was the Management–Union Committee on EAP Projects, hereafter referred to as the *Joint Committee*. As its name implies, it consisted of senior management appointees from government and senior union appointees from the Ontario Public Service Employee's Union.

The Joint Committee took the view that it would be necessary to carry out surveys of the need for EAPs in specific workplaces where interest was being expressed so that the key actors (management and labor) could have a relatively objective basis upon which to judge the prevalence of problems that the proposed program was meant to address. In addition, the results of such prevalence surveys could serve as baseline data against which the impact of newly introduced programs could be evaluated.

It was decided that the surveys would have to encompass more than the assessment of alcohol-related problems since it was generally known and accepted that voluntary EAPs tended to generate clients with a wide variety of difficulties and that most of them interfered with personal functioning on and off the job sooner or later. The Joint Committee therefore accepted a proposal from the authors to the effect that surveys should include two major components under the heading of a "Brief Mental Health Inventory," namely a measure of distress and a measure of alcohol involvement. The prevalence of distress (to be defined shortly) was taken to be a proxy for the variety of circumstances and conditions that might form the basis of presenting problems in EAP and a rationale for the development of preventive programming (health promotion).

Survey Method

The Brief Mental Health Inventory was administered in four separate sites under the auspices of local joint management–union committees. In every case, however, the actual administration of the questionnaire was overseen by the Joint Committee's consultant (L. M. Poudrier). For the most part, the inventory was administered in large group settings where union officials could insure confidentiality and anonymity. Prior to administration, however, employees were informed by letter of the survey's imminence and of its objectives. These letters were usually sent by local joint committees or by management with the committee's endorsement. It will be apparent from the foregoing that the survey process was part of a larger one involving the development of closer local working relations between labor and management.

In all cases, the survey was administered during working hours on a census rather than a random basis to avoid union concerns about perceived discrimination in selective surveys. Only age and sex identifiers were used since the question of "occupation" aroused fears about potential identification of individuals in sites where only a few incumbents of certain positions were to be found.

The Survey Instrument

The Brief Mental Health Inventory consisted of the Derogatis SCL-90-R (Derogatis 1977) and the Mortimer–Filkins (M–F) questionnaire. The prior use of the Mortimer–Filkins by the first author and others has been described elsewhere (Shain and Groeneveld 1980). *The SCL-90-R (Derogatis 1977)* is a ninety-item self-report symptom inventory developed from the Hopkins Symptom Checklist. Its history is traceable to the Cornell Medical Index. Respondents are rated on nine primary symptom dimensions as follows: somatization, obsessive-compulsiveness, interpersonal sensitivity, depression, anxiety, hostility, phobic anxiety, paranoid ideation, and psychoticism. These dimensions are represented by subscales. In addition, three global indices can be generated: The Global Severity Index (GSI), the Positive Symptom Distress

Index (PSDI), and the Positive Symptom Total (PST). The first of these, the GSI, was used in this study. Scores on the GSI are used here as a measure of general distress.

The symptom dimensions have been shown to have an acceptably high internal consistency, ranging from .77 to .90 (coefficient alpha). Test–retest reliability results have been similarly encouraging, coefficients ranging from .78 to .90 (Derogatis, Rickels, and Rock 1976).

The first five dimensions of the SCL-90-R (which include anxiety and depression) were found to have a high level of factorial invariance across social class and psychiatric diagnosis (Derogatis et al. 1971; 1972). Using an invariance coefficient developed by Pinneau and Newhouse (1964), Derogatis and Cleary (1977) also showed high levels of factorial invariance across sexes. Clearly, it is important to know that the integrity of the structure of the symptom dimensions is maintained in such comparisons. Such findings contribute to arguments about the utility of the SCL-90-R in comparisons of the symptomatology of different work forces.

The validity of the SCL-90-R has received a respectable amount of attention given its quite recent origins. It has fared quite well in tests of its convergent validity (with, for example, the MMPI and the Middlesex Hospital questionnaire). Its discriminative validity appears to be quite solid in that the profiles of a number of different populations (both clinical and normal) can be readily distinguished. In addition, three major normative samples exist: 1,002 heterogeneous psychiatric outpatients, 974 nonpatient normals, and 112 adolescent psychiatric outpatients. While it may be possible to deduce construct validity from the foregoing types of data, the SCL-90-R has been factor analyzed against a hypothetical framework. This involved a comparison of the presumptive dimensional structures of the symptoms with an actual factor structure. The match, reported in Derogatis (1977), tends to strongly support the integrity of the hypothesized structure. The sample used for this procedure were the 1,002 psychiatric outpatients referred to above. However, it is clear that the dimensions are not orthogonal. One might expect a certain degree of cross-loading of individual symptoms on more than one factor. This is indeed what one observes in the case of eight of the ninety symptoms that load on more than one factor.

Among the profiles of various populations tested with the SCL-90-R is one of 110 male applicants to the Hopkins Alcohol Rehabilitation Unit (Derogatis 1977). Another profile of an employed population (154 New York Telephone Co. employees) who self-selected themselves for a course on stress management has been developed by Carrington and associates (1980).

Thus, sufficient data have been accumulated using the SCL-90-R (or its close cousins) to provide an interpretive base for analysis of findings from the survey reported here.

The Mortimer–Filkins questionnaire is a self-report device that originally formed part of the "Court Procedures for Identifying Problem Drinkers" developed by the University of Michigan Highway Safety Research Institute (Kerlans et al. 1971). It was intended to be used in conjunction with interview data and information about previous criminal involvement to aid in forming a judgment about the likelihood that a person convicted of an impaired driving offense was an alcoholic. The assignment of scarce treatment resources could then be made on the basis of such judgments. In fact, the questionnaire alone can be used to form these judgments but the results are not quite as accurate as when they are compared with findings from the other two sources of data. However, the use of the interview and information about alcohol-related criminal involvement (particularly, DWI) in the past as well as BAC at time of arrest can be seen as providing a type of construct or at least concurrent validity for the questionnaire results when the various sources are compared. Indeed, the original point-biserial correlations between the questionnaire results and criterion group membership based on the other sources of information were .83 for males and .80 for females (Mortimer et al. 1971). In a later study of 78 impaired drivers in Oshawa, Ontario, Ennis and Vingilis (1981) reported that the questionnaire alone successfully classified 67 percent of respondents as problem drinkers when the external criterion (CRIT) defined problem drinking in relation to a BAC at time of arrest of at least 0.20 percent, or two or more prior DWI offenses, or three or more other alcohol-related offenses. Another 22 percent who were classified by the CRIT as problem drinkers were classified as "presumptive problem drinkers" by the questionnaire. According to the CRIT, therefore, only 11 percent who were problem drinkers or presumptive problem drinkers were misclassified by the questionnaire as social or normal drinkers. If we consider only the 67 percent who were classified as problem drinkers by both questionnaire and CRIT, we see that the convergence parallels the covariance found by Mortimer and associates (1971) between the questionnaire and criterion group membership (.83 for males translating as approximately 69 percent covariance; .80 for females as 64 percent). The convergence of classification with that produced by external criteria increases, however, when the goal is to distinguish social or normal drinkers from both problem drinkers and presumptive problem drinkers. As shown above, this results in only an 11 percent misclassification by the questionnaire.

In the original test validation trials, it was found that if one wished on the basis of questionnaire scores to classify 99 percent of the 192 problem drinkers correctly ("correctness" being judged by known membership of a problem-drinking group—such as patients on an alcoholism ward) then approximately 10 percent of a control group would also be classified as alcoholics. This finding is discussed as though the 10 percent were *misclassified*. Consequently, we see the cutting points eventually employed being adjusted so that 95 percent of problem drinkers are correctly identified while only 5 percent of the controls are thus misclassified. The assumption throughout appears to have been that the control group were all normal or social drinkers. However, the control group contained seventy-two firemen, a number of married students and faculty members from a university, people collecting unemployment benefits or looking for jobs, as well as people from synagogues and churches. Thus, it is not improbable, indeed it is most likely, that the controls contained at least 5 percent problem drinkers, based on normal population estimates (e.g., Cahalan and Room 1974). While it is still likely that some misclassifications occured, it seems wrongheaded to assume that the majority of the 5 percent were actually misclassifications. Consequently, it seems likely that the M–F questionnaire cutting points recommended in the manual (Mortimer et al. 1971) are adjusted to produce considerably more false negatives than false positives. Thus, results based upon M–F scores—at least with regard to designation of problem drinkers—are likely to be quite conservative.

An additional reported characteristic of the questionnaire from the original trials relates to its split-half reliability coefficient, which is .904. Jacobsen (1975) has suggested the potential use of the M–F questionnaire in survey work. Its reported construct, convergent and discriminative validity combined with its low face validity as a "drinking" inventory make it attractive in cases where low threat is a desirable characteristic. For this reason it is preferable to the Michigan Alcoholism Screening Test (MAST). Conversely, the M–F questionnaire does not provide a differential diagnosis or an assessment of severity or progression, as indicated by Ennis (1977). These are functions probably better served by the multidimensional MAST, as shown by Skinner (1979). However, the present research further develops our knowledge about the dimensionality of the M–F questionnaire in the context of its relationship to scores on the SCL-90-R. In addition, studies of these relationships yield estimates of populations at risk generated from convergences of findings from the M–F and SCL-90-R questionnaires, thus adding to the utility of both.

The M–F questionnaire has been used for purposes similar to those reported here in several sites in Ontario where the objective was to provide an estimate of the population at risk for drinking-related problems and the (probable) consequent need for an EAP. Three of these studies have been reported in Shain and Groeneveld (1980). The samples have included psychiatric institution staff, armed forces personnel, and employees at a large Ontario ministry. Results showed considerable variation in populations at risk across sites. An approximate estimate of within-site stability was also obtained from a readministration of the M–F questionnaire in a ministry four years after the first administration (Gray and Poudrier 1982). While this was not a cohort study, the proportions of problem drinkers to presumptive problem drinkers and social or normal drinkers were almost identical on both test occasions. While this is clearly not an example of test–retest reliability, it may bear some degree of relationship to the effect described by Cook and Campbell (1979) as "cyclical turnover." This simply refers to the phenomenon whereby the profiles of institutional members remain stable over time even though individual members continue to change by virtue of turnover. This effect is maintained, presumably, by a constancy in recruitment, hiring, training, and management practices.

Survey Sites

Of the four major sites, three were institutions, the fourth a provincial government ministry. In tables 3–1, 3–2, 3–3, 3–4, and 3–5 it will be seen that these are Sites A, B, C, and E. Site D was not part of the surveys conducted by the Joint Committee but it is included here for comparative purposes. The circumstances under which Site D was surveyed differ from the other four in that a local joint committee was not involved, although the intent of management was to provide data for determining the need for an EAP.

In the case of the four principal sites, the invitation to conduct the surveys came from local joint committees to the provincial Joint Committee. In all four cases it was known that the work forces in question were more than usually troubled even by institutional standards. It is commonplace to observe that staff in psychiatric hospitals, jails, and correctional institutions are exposed to highly stressful working conditions, but the four sites examined here were considered to be particularly difficult working environments.

The return rates for the five sites varied, with three of them exceeding 80 percent and one falling below 50 percent: Site A, 81 percent (large psychiatric hospital); Site B, 86 percent (medium-sized correctional institution for juveniles); Site C, 88 percent (large city jail); Site D, 69 percent (municipal government employees); Site E, 45 percent (head office employees, provincial ministry). The return rates in all cases except Site D were functions of the degree to which local joint committees were able to generate enthusiasm for the surveys. In the case of Site E, it is difficult to conclude to what extent the results can be generalized to the work force from which the 45 percent sample selected itself. However, the results still serve as a partial indicator of the prevalence of problems even though clearly they do not provide a complete picture.

Results

The SCL-90-R Global Severity Index

The scores on the GSI are shown for males in table 3–1 and for females in table 3–2. In these tables, the GSI scores are grouped in such a way as to show their relationship to the norm data provided in the Derogatis manual for nonpatient populations—that is, people selected at random from a large metropolitan center (Derogatis 1977). We employed this method of illustrating the data rather than simply reporting mean scores for each site because the scores were sometimes abnormally distributed, thus making a comparison of means somewhat misleading in itself.

The most striking aspect of the distress scores is that they were heavily skewed toward what Derogatis refers to as the clinical range since scores in these ranges tend to characterize patients in residential psychiatric care. In tables 3–1 and 3–2 these clinical range scores appear as the fourth group. The most noticeable populations in this regard were male and female staff at Site A and male staff at Sites B, C, and E.

A comparison of our GSI distress scores with those of Carrington and associates obtained in their New York Telephone Co. study shows that several of our samples resemble people who volunteered for stress management courses at some mid-point during those courses (Carrington et al. 1980). Similarly, readers will find that the participants in the stress management courses reported in chapter 6 of this book scored higher than employees in the sites considered here before their training but lower afterwards.

Clearly, the municipal government employees (Site D) were the most "normal" in comparison with the Derogatis nonpatient populations. It will be recalled that most employees in this group were not employed in institutions. However, even in this sample it turned out that a subgroup of the women were staff in a nursing home. It was possibly the scores of this subgroup that skewed the female mean toward the clinical end of the range in this case. Males in this group (Site D) most closely resemble the Derogatis nonpatient males.

It is interesting that distress scores in Site E (the provincial ministry) were elevated to levels comparable with those found among institutional staff, particularly among females (see groups 3 and 4). However, we should view these results with considerable caution since they are based on a 45 percent return rate. The important point about results such as these from a practical point of view is that they demonstrate the existence of a needy group within the work force in question. These surveys produce at least *minimum estimates* of populations at risk. Clearly, however, the planning of any intervention according to survey results is on much firmer ground when the return rate is in the 80 percent range. With lower return rates, the probability is that the need for intervention will be underestimated.

While comparisons of distress levels based on GSI scores across different samples are instructive, it is important also to place all these estimates within a more general interpretive framework. It has been suggested, for example, that at any one time up to 15 percent of the population is actually in need of some form of mental health treatment while as many as 25 percent may be suffering from mild to moderate depression, anxiety, and other forms of emotional distress (President's Commission on Mental Health 1978; see also Wei-

Table 3–1
Derogatis Global Severity Index (GSI)[a] Scores in Five Survey Sites: Males

GSI Scores	Site A: Large Psychiatric Hospital		Site B: Medium-Sized Correctional Inst. for Juveniles		Site C: Large City Jail		Site D: Municipal Gov't. Employees		Site E: Head Office Employees, Provincial Ministry		Total	
	n	%	n	%	n	%	n	%	n	%	n	%
1. 0–.17	47	(28.8)	9	(15.0)	35	(27.3)	19	(47.5)	33	(29.7)	143	(28.5)
2. .18–.45[b]	47	(28.8)	23	(38.3)	41	(32.3)	14	(35.0)	40	(36.0)	165	(32.9)
3. .46–.87[c]	22	(13.5)	18	(30.0)	23	(18.0)	3	(7.5)	21	(18.9)	87	(17.3)
4. ≥ .88[d]	47	(28.8)	10	(16.7)	29	(22.7)	4	(10.0)	17	(15.3)	107	(21.3)
Total	163	(100)	60	(100)	128	(100)	40	(100)	111	(100)	502	(100)

[a]Derogatis (1977).
[b]Relative to the norm for nonpatient males (Derogatis 1977), these scores fall between the 50th and 83rd percentiles.
[c]Relative to the norm for nonpatient males (Derogatis 1977), these scores fall between the 84th and 98th percentiles.
[d]Relative to the same norm, these scores fall beyond the 98th percentile.

Table 3–2
Derogatis Global Severity Index (GSI)[a] Scores in Five Survey Sites: Females

GSI Scores	Site A: Large Psychiatric Hospital		Site B: Medium-Sized Correctional Inst. for Juveniles		Site C: Large City Jail		Site D: Municipal Gov't. Employees		Site E: Head Office Employees, Provincial Ministry		Total	
	n	%	n	%	n	%	n	%	n	%	n	%
1. 0–.25	88	(28.0)	14	(45.2)	15	(36.6)	32	(33.0)	48	(23.9)	197	(28.8)
2. .26–.59[b]	89	(28.3)	12	(38.7)	12	(29.3)	28	(28.9)	77	(38.9)	218	(31.9)
3. .60–1.48[c]	87	(27.7)	4	(12.9)	13	(31.7)	30	(30.9)	58	(28.9)	192	(28.1)
4. ≥ 1.49[d]	50	(15.9)	1	(3.2)	1	(2.4)	7	(7.2)	18	(9.0)	77	(11.2)
Total	314	(100)	31	(100)	41	(100)	97	(100)	201	(100)	684	(100)

[a]Derogatis (1977).
[b]Relative to the norm for nonpatient females (Derogatis 1977), these scores fall between the 50th and 83rd percentiles.
[c]Relative to the norm for nonpatient females (Derogatis 1977), these scores fall between the 84th and 98th percentiles.
[d]Relative to the same norm, these scores fall beyond the 98th percentile.

ner, Akabas, and Sommer 1973). This seemingly high estimate finds some confirmation in the data presented here. If we take the 84th percentile of scores on the GSI relative to the Derogatis nonpatient norms as indicating the point at or above which people are "seriously handicapped by emotional problems" (temporarily or otherwise) we can see that "normally" 16 percent of the population would fall above this point. In our own samples reported here we find that only the males in Site D and females in Site B exhibit scores on the GSI that hover within the normal range (17.5 and 16.1 percent respectively falling above the 84th percentile). In contrast, for all sites combined, 39.3 percent of all female respondents and 38.6 percent of all males had scores at or above the 84th percentile. This represents more than double the expected rate. Remember, however, that four of the five sites had been selected because they were known trouble spots. Consequently, one would expect to see evidence of this perception in the form of higher than normal distress scores. Relative to the estimate provided by the President's Commission on Mental Health (1978) the scores from the present samples are still extremely high, but it should be kept in mind that the criteria for such comparisons are lacking. By presenting such a comparison at all, we are seeking merely to place our own figures within a context in which the term "serious handicap due to emotional problems" has an intuitive rather than empirical meaning.

The Mortimer–Filkins Questionnaire

Scores are shown for males in table 3–3 and for females in table 3–4 and are grouped according to risk levels established from the manual (Kerlans et al. 1971). Risk refers to the probability that, based on these scores alone, the respondent is a problem drinker. High risk, therefore, indicates a very high probability, medium risk a fairly high probability, and low risk a correspondingly low probability. People classified in the medium-risk category are also referred to as presumptive problem drinkers (Kerlans et al. 1971). This language is intended to convey the idea that further evidence is required to either confirm or disconfirm the likelihood that the respondent is a problem drinker.

It can be seen that among males, the highest risk category is occupied by a low of zero percent (Site D) and a high of over 12 percent (Site A). Among females, the corresponding figures are zero (Sites B and C) and 10.2 percent (Site A). However, Sites D and E each claim only 1 percent of females in the highest category.

In previous reports of problem drinking among employed populations using the M–F, it was shown that the highest risk category was usually occupied by between 1 and 2 percent of the work force, while the medium-risk group usually comprised between 9 and 12 percent (Shain and Groeneveld

Table 3–3
Mortimer–Fillkins[a] Score Groups in Five Survey Sites: Males

Scores	Site A: Large Psychiatric Hospital		Site B: Medium-Sized Correctional Inst. for Juveniles		Site C: Large City Jail		Site D: Municipal Gov't. Employees		Site E: Head Office Employees, Provincial Ministry		Total	
	n	*%*	*n*	*%*	*n*	*%*	*n*	*%*	*n*	*%*	*n*	*%*
0–15 (Low Risk)	113	(69.3)	51	(85)	98	(76.6)	38	(95)	103	(92.8)	403	(80.3)
16–24 (Medium Risk)	30	(18.4)	7	(11.7)	22	(17.2)	2	(5)	7	(6.3)	68	(13.5)
25 + (High Risk)	20	(12.3)	2	(3.3)	8	(6.3)	—	—	1	(0.9)	31	(6.2)
Total	163	(100)	60	(100)	128	(100)	40	(100)	111	(100)	502	(100)

[a]Kerlans and associates (1971).

Table 3–4
Mortimer–Filkins[a] Score Groups in Five Survey Sites: Females

Scores	Site A: Large Psychiatric Hospital		Site B: Medium-Sized Correctional Inst. for Juveniles		Site C: Large City Jail		Site D: Municipal Gov't. Employees		Site E: Head Office Employees, Provincial Ministry		Total	
	n	%	n	%	n	%	n	%	n	%	n	%
0–15 (Low Risk)	246	(78.3)	27	(87.1)	35	(85.4)	88	(90.7)	182	(90.5)	578	(84.5)
16–24 (Medium Risk)	36	(11.5)	4	(12.9)	6	(14.6)	8	(8.2)	17	(8.5)	71	(10.4)
25 + (High Risk)	32	(10.2)	—	—	—	—	1	(1.0)	2	(1.0)	35	(5.1)
Total	314	(100)	31	(100)	41	(100)	97	(100)	201	(100)	684	(100)

[a]Kerlans and associates (1971).

1980). Thus, for both males and females, Sites A and C (both institutions) were unusual in the high frequency of problem or presumptive problem drinkers.

Dual Risk Estimates

Dual risk estimates can be produced by calculating the number of people who scored in the highest risk categories on both the SCL-90-R GSI and the M–F. The results can be found in table 3–5.

The two sets of estimates vary in their conservatism. Estimate A requires that qualifying respondents should be placed both in the highest Mortimer–Filkins risk group (problem drinker) and in either of the two highest GSI categories (scores at or above the 84th percentile in the Derogatis norm sample of nonpatients). Estimate B allows respondents to qualify for "dual risk" designation if they were either problem drinkers (high risk) or presumptive problem drinkers (medium risk) and scored above the 84th percentile relative to the nonpatient norm for the GSI. Thus, by adjusting the criteria for inclusion, one may derive either a more conservative or a more liberal estimate of the population at risk. It is worth noting that GSI scores and the two subscales of the M–F (alcohol-specific and distress) are highly correlated among the samples considered in this chapter. While it might be expected that the GSI and the distress subscale of the M–F would be related, it is somewhat surprising that the GSI and the alcohol-specific subscale of the M–F are related to the extent of 0.66 on the average (table 3–6).

The high correlations shown in table 3–6 illustrate the extent to which high scores on the GSI predict high scores on the M–F and vice versa. According to scattergram plots of these correlations, it is evident too that the higher the scores, the closer the relationships.

Interpretation

The data from the Brief Mental Health Inventory provided a basis upon which decision makers in individual sites could establish, taking into account other types of information as well, the extent of need not only for employee assistance but also for health promotion programs, at least those that have clear mental health components such as stress management courses. The problem, of course, lies in the interpretation of the facts that tumble over each other in an effort to portray a coherent picture of mental health or the lack of it. Interpretation, naturally, hinges upon individual perspectives. How bad does distress have to be, and how prevalent, before action should be taken? How serious is "presumptive problem drinking"? When someone is both highly distressed and a problem drinker (not an unexpected combina-

Table 3–5
Dual Risk Estimates Based on SCL-90-R Global Severity Index Scores[a] and Mortimer–Filkins[b] Scores

Estimate A: Percentage over the 84th nonpatient norm percentile (GSI) and also in the problem drinker group (Mortimer–Filkins)

	Males		Females	
Sites	*n*	%	*n*	%
A	20	(12.3)	31	(9.8)
B	2	(3.3)	0	
C	6	(4.7)	0	
D	0		0	
E	2	(1.8)	4	(2.0)

Estimate B: Percentage over the 84th nonpatient norm percentile (GSI) and also in the presumptive problem or problem drinker group (Mortimer–Filkins)

	Males		Females	
Sites	*n*	%	*n*	%
A	38	(23.3)	60	(19.1)
B	7	(11.6)	1	(3.2)
C	19	(14.8)	3	(7.3)
D	1	(2.5)	6	(6.2)
E	3	(2.7)	7	(3.5)

[a]Derogatis (1977).
[b]Kerlans and associates (1971).

tion), what response is most effective? It is worth emphasizing that "distress" in the sense conveyed by the GSI of the SCL-90-R is no empty label: it is a superscore derived from items that measure a wide variety of conditions—anxiety, depression, interpersonal sensitivity, obsessive-compulsiveness, hostility, phobic anxiety, paranoid ideation, somatization, and psychoticism. When the GSI is as elevated as it frequently was in the five sites examined here, it is because the scores on these subscales are elevated too, though not all to the same extent. Clearly, distress is a useful term to describe a wide range of symptomatology across a wide range of mental health dimensions. A distress superscore (GSI) indicating that an employed person is manifesting the symptoms of someone who could easily be found in a psychiatric hospital is difficult to ignore. It is probably not melodramatic to say that people in such conditions are walking time bombs. Their need is urgent. Lesser extremes, however, are more difficult to interpret. It is possible to argue, sometimes rightly no doubt, that such distress may be "situational" and that it will

Table 3–6

Pearson Correlations between the Derogatis Global Severity Index (GSI)[a]
and Mortimer–Filkins (M–F) Subscales[b]

Correlations	Site A: Large Psychiatric Hospital n = 464	Site B: Medium-Sized Correctional Inst. for Juveniles n = 92	Site C: Large City Jail n = 171	Site D: Municipal Gov't. Employees n = 143	Site E: Head Office Employees, Provincial Ministry n = 321
GSI with M–F Score 1 (Alcohol)					
Coefficient	0.82	0.65	0.67	0.59	0.57
Significance	.001	.001	.001	.001	.001
GSI with M–F Score 2 (Distress)					
Coefficient	0.85	0.82	0.83	0.82	0.68
Significance	.001	.001	.001	.001	.001

[a]Derogatis (1977).
[b]Kerlans and associates (1971).

pass when the crisis engendering it passes. It is apparently true that even very high distress levels are amenable to nonheroic interventions such as stress management programs (see, for example, chapter 6, and also Carrington et al. 1980). Consequently, a high GSI score does not necessarily nor even probably mean that the sufferer is "mentally ill" in the sense of being diagnosable according to the DSM III (American Psychiatric Assn. 1980). Of course, within this group of high scorers is a subgroup of unknown size whose SCL-90-R profiles probably would present preliminary evidence of diagnosable conditions.

Indeed, in each of the four principal sites where committees existed, the results were interpreted in different ways. This is not surprising since no common framework for interpreting results below the clinical or "red light zone" exists. Even at this level there is room for disagreement among professional diagnosticians and care givers. In some instances the distress scores were considered important in their own right; in others they were seen as important only in confirming the proportion (though, of course, not the identities) of people who could be considered problem drinkers in immediate need of attention. This is perhaps as it should be, since there is no objective definition of need for intervention—in the same way that there is little consistency in diagnosis—except in the most extreme of circumstances where the preponderance of evidence suggests that a crisis exists.

While the results of the present study are part of a decision-making and fact-finding process, they do not provide data sufficient for the formation of conclusions on the etiology of the observed conditions. Nonetheless, it is incumbent on the present writers to offer some hypotheses about the genesis of the mental health profiles that have been described in this report, for the purpose of provoking further discussion and research. We draw upon a variety of theoretical frameworks that are not intended to be mutually exclusive. Indeed, the processes to which they refer may have additive, multiplicative, or otherwise interactive effects.

1. It is a commonplace observation among students of remedial institutions that the frequent juxtaposition of custodial/disciplinary objectives with treatment/rehabilitative objectives can set up stresses and strains in organizations such that effectiveness is compromised and staff are placed in a state of perpetual role conflict (see Goffman 1961; Street, Vinter, and Perrow 1966; Piliavin 1966; Shain 1976; Cheek and DiStefano Miller 1982 for examples). This inconsistency in objectives is often quite acute in correctional and psychiatric institutions where transitions between custodial regimes and more rehabilitative regimes are most likely to be encountered. The transition is more likely to be perceived as confusion than evolution by line staff who may be expected at one moment to engage their charges in a meaningful therapeutic relationship and at another moment to lock them up. While the apparent contradiction in behaviors may have a rational basis it is difficult to communicate this either to inmates, patients, residents, clients, or staff.

Training of staff in methods of rehabilitative management or treatment can even exacerbate the stressfulness of role conflict or role ambiguity. Training may be perceived as a threat if it implies that previous staff attitudes and behavior were wrongheaded. This threat can produce premature rejection of new ideas accompanied by anxiety. It can also produce paralysis, withdrawal, and depression (see Cloward et al 1960 for discussion of these responses). Obviously, none of these reactions add to the effective management of inmates or patients let alone to their rehabilitation. Indeed residents tend to perceive their keepers' reactive behavior as introducing even more uncertainty and lack of direction into their own lives. People require some defense against this sense of uncertainty and meaninglessness. One of the most common defenses in correctional settings appears in the formation of what has traditionally been called the "inmate culture" (see Schrag 1954; Clemmer 1958; McCorkle and Korn 1954; Sykes and Messinger 1970 for examples). The inmate culture is typified by the group espousal of norms that are in opposition to treatment/rehabilitation goals. As inmate group cohesion consolidates around delinquent or deviant values, so the withdrawal and hopelessness of staff may increase. Thus a looping chain reaction is set off with dire consequences to the mental health of staff and to the chances that inmates will shed or modify their deviant code of behavior. There is reason to believe that this reactive group formation can occur in any custodial residential situation (see Goffman 1961, for example).

2. It is likely that at some point during a transitional regime a sizable body of staff will exist whose basic values are or appear to be at odds with those underlying the new goals of the institution. This may be ultimately resolved through transfers, retirement, and retraining. However, the limbo-like situation can be perpetuated by recruitment practices that do not reflect the needs of the emerging institution, for example, there is little use in hiring authoritarian staff to serve treatment-oriented goals. While this is an extreme example, various lesser forms of it are frequently encountered.

Both explanations so far have been couched in terms of problems that surround organizational change—planned or otherwise. They assume, therefore, that solutions to the problems we have noted are primarily organizational or structural in nature: there is something that senior management, in consultation with staff, can do to correct the situation by changing the way the place is run. Help from outside may be required, but the focus for action is the organization itself, not the psyches of the staff as such. (This issue is explored in greater depth in chapters 9 and 10.) Other hypotheses make different assumptions, and they follow.

3. Front-line work with delinquent or disturbed residents is intrinsically stressful, even when all the resources of the institution are supportive of individual staff effort. Much energy and skill are required to reach people whose problems have led to their being institutionalized. This work is frequently unrewarding and frustrating. Its course is unpredictable and the balance between success and failure is a fine one, requiring of the staff that they have a high tolerance for ambiguity. In and of itself this kind of work may generate stress and burnout among all but the most resilient. Without extraordinary organizational support, increased involvement with inmates and their problems can lead to demoralization or burnout. Withdrawal and "distancing" produce short-term relief but probably long-term depression (see, for example, Cheek and DiStefano Miller 1982). Of course, this type of stress is not peculiar to residential situations, but it is likely to be exacerbated within them.

4. Staff may bring problems with them to the workplace. In some cases this must surely be true. The question becomes whether the mental health problems described earlier in this chapter can be accounted for by the prevalence of such problems among those people who are attracted to the work of institutions before they are actually employed in these settings. At this point, it seems necessary to invoke the vague but intuitively appealing concept of "job-person fit" (see McMichael 1978; Van Harrison 1978 for examples). When a person looks for a job, he or she ideally tries to match his or her characteristics with the job in question. In practice this is often, perhaps usually, an imperfect match. However, the mismatch can be made all the worse if the employer changes the game plan by, for example, switching from one set of organizational objectives to another. The resulting tension

can then be as much attributed to organizational precipitation as to individual predisposition.

5. The relationship between distress and alcohol involvement noted earlier in this chapter suggests a cyclic development of drinking problems and more general mental health problems in which alcohol is used to reduce tension and to alter the disagreeable moods associated with high distress. This self-medication approach to the relief of distress may work for some people, for limited periods of time and for transient conditions, but the potential for development of dependence upon alcohol under these circumstances is very high, whether drinking actually relieves psychic discomfort or is simply believed to do so (Cappell 1975). Given the data in the present study, it is virtually impossible to determine whether heavy drinking in itself acted as a stressor that led to the development of other mental health problems as opposed to the reverse process where distress led to drinking. Beyond a certain point, however, the issue becomes academic, since excessive drinking and high distress tend to be negatively synergistic.

Interventions

A discussion of interpretation leads to the practical implications of the various theoretical and even philosophical perspectives that can be brought to bear on the data at hand. As noted at the beginning of the chapter, the surveys that produced the data presented here were intended to assist in determining the extent of need for voluntary EAPs. Thus, an assumption underpinned much of the work and tended to prejudice the nature of discussions concerning its practical implications. The assumption was that the problems brought to light (or confirmed) by the surveys would be amenable to solutions found within the framework of EAPs. Accordingly, the implicit model of intervention was one in which *treatment* (as provided by clinicians) is provided to *clients* (quasi-patients), usually on a *one-to-one* basis. This implication tended to prevail even in relation to the very wide range of problems that local joint committees expected to emerge in their home sites once the EAP was introduced. This, of course, is not a unique implication: it tends to prevail wherever EAPs are introduced. However, it will have become apparent in reviewing the various interpretations of the data from broader theoretical perspectives that we ought not to allow the assumption to stand and that several other implications can be derived, some of which require a move beyond the clinical framework within which EAPs are usually considered.

The very size of the populations at risk indicated by the surveys suggests that EAP cannot be the sole response to problems which are so prevalent. If we return to our "dual risk" estimates shown in table 3–5, it can be appreciated with reference to estimate B that even these projected numbers of people in need would tax the resources of most EAPs. If we move beyond this method of estimation and return to estimates of the population at risk based on either the distress scores (SCL-90-R) or the alcoholism-risk scores (M–F) we find that the numbers are so large that we must either reject them as being based on too liberal a definition of risk and need or we must allow the possibility that other interventions might be relevant for part of the identified group.

Our position is that the definition of risk and need is not too liberal but rather that the *level* of risk and need indicated by the greater than normal but less than clinical range distress scores often suggests a secondary rather than tertiary intervention. Secondary-level interventions are those that assume the individual to be still in some control of his or her life and are directed toward the strengthening and development of skills as well as at the remediation of problems. The stress management programs that are evaluated in chapter 6 are prime examples of these interventions. To a lesser extent, the healthy life style course considered in chapter 5 is also of relevance at a secondary level, although its greatest strength may lie in the primary prevention area. Chapter 2 has provided an extensive rationale for taking the needs of what we might call the "moderate risk" group very seriously. Not only are people in this category a source of loss to themselves, their families, and their employers, but also it is from their ranks that candidates for EAPs eventually spring. In chapter 7, arguments are developed to the effect that an extension and blending of traditional EAPs and HPPs is likely to be not only more effective in reaching the moderate risk group but cost-efficient too. For present purposes, we are proposing that typically the primary- and secondary-level interventions referred to in the foregoing be considered under the heading of HPPs.

Useful though programmatic interventions of the EAP/HPP type undoubtedly are, we must draw attention to the fact that much of the hypothesis and speculation considered under the head "Interpretation" concerned itself with the *structural* origins of distress and problem drinking. If this point of view has any legitimacy at all (chapter 8 is a detailed elaboration of it) it implies that the level of distress in a given work force is at least partly a function of the way in which the organization is managed. Consequently, modification of the socio-technical structure of work will have a discernible impact upon the mental health of workers. As chapter 10 shows, however, tinkering with structure can affect mental health in a variety of ways, both good and bad. Nevertheless, the fact remains and can be well documented (see chapter 9) that the organization of work always has an impact upon mental health: it is never a neutral set of conditions. Consequently, it makes little sense—particularly under extreme circumstances—to introduce programmatic aids such as EAP and HPP when defects in the socio-technical structure of the organization render them virtually impotent. One might argue that even in "bad" environments it would be useful to have programmatic aids simply to provide some comfort, relief, and first-aid to employees. For example, stress management courses might be seen as providing self-defense techniques for participants. However, much of the potential impact of these programs can be obliterated by the way in which they are introduced and presented to the work force. If they are seen as bandaids or as methods by which management seeks to avoid its responsibilities for providing a work environment that is not superfluously stressful, they are likely to be treated with suspicion and be poorly attended. Further, the benefits of stress management training can be effectively destroyed or at least heavily diluted by returning participants to environments that are inimical to their mental health.

Our view is that when surveys indicate that high distress is running at double the expected rate among the work force, the management and union should take a close look at how the work of their organization is being undertaken. It is extremely unlikely (though it would be in itself a structural problem) that recruitment practices account for distress rates of this order. More likely is the explanation that the organization of work is placing unnecessarily stressful pressures upon employees in terms of, for example, work overload, role conflict, ambiguity and confusion, lack of discretionary authority (or too much of it), job fragmentation (or over-extensiveness), poor communication channels, lack of scheduling flexibility, and so forth. This may not be a deliberate policy—indeed, it is unlikely to be since it is associated with low organizational effectiveness—but it is a policy by default. Recall once more that four of the five sites considered here were selected because they were known to be trouble spots. This in itself implies that other psychiatric hospitals, jails, correctional centers, and government departments were not so troubled and yet they were doing the same kind of work within their own categories. This phenomenon points to an essential principle, which is that the same sorts of people doing the same kinds of work but employed under different organizational conditions can manifest wide variations in their mental health.

In conclusion, then, we argue that amelioration of workers' mental health must be considered from both structural and programmatic points of view. Considerable care must be taken to avoid making the assumption that all mental health problems are amenable to programmatic interventions. While the latter are powerful allies in attempts to safeguard and restore the mental health of workers, they become no more than solutions in search of problems to fit them if we do not keep open minds about the range of structural shortcomings that can subvert the most promising of programmatic interventions.

4
The Prevalence of Health Promotion Activities in the Workplace

In the first three chapters we reviewed the rationale for developing health promotion programs as part of the gamut of approaches that are required to address the health needs of the whole work force, thereby adding a preventive component to the services already available through broad-based EAPS. In practice, HPPs are generally not planned and carried out in coordination with EAPs, although increasingly both types of intervention are found in the same organizations. In this chapter, we examine the evidence concerning the prevalence of HPPs both in their own right and in relation to EAPs. The evidence to date is somewhat fragmentary but worthy of consideration since it shows a fairly wide distribution of HPPs, suggesting that for good or ill, these forms of intervention are a force to be reckoned with as methods are sought to develop comprehensive approaches to the mental and physical well-being of workers.

We shall discuss the data in two parts: first, that which deals with HPPs in general; and second, that which deals with HPPs in the context of EAPs.

General Studies of Health Promotion Prevalence

Fielding and Breslow (1983) conducted a study of HPPs in California. They noted that most studies thus far had focused on large organizations, among which it was typically found that between 30 and 40 percent had been involved in stress management programs and about 25 percent in fitness programs. In their own study, Fielding and Breslow drew a random sample of private employers with more than 100 employees. They ended up with a sample of 424 organizations, identifying in each the person who was designated as most responsible for health promotion and requesting a 30-minute, structured interview. A similar study was carried out in Ontario by Danielson and Danielson (1980) on behalf of the then Ministry of Culture and Recreation, now the Ministry of Tourism and Recreation (Sports and Fitness Branch). This study was partially replicated in 1984 by Dean, Reid, and

Gzowski (1984) on behalf of the same ministry. The results of the 1980 and 1984 Ontario studies were quite similar. Because the 1980 study was reported in a way that permits comparison with Fielding and Breslow's study, the following discussion emphasizes this rather than the 1984 report, although readers are asked to note that the two sets of data portray much the same picture except where specifically stated.

The Danielson and Danielson study was also based upon a random sample, in this case yielding a 70 percent return rate of 680 organizations employing more than 50 people. Fielding and Breslow reported a return rate of 83 percent. The Danielsons employed mail-out questionnaires in comparison with the Fielding and Breslow interview method. In the former case, personnel directors rather than the person most responsible for health promotion were the intended targets of the survey.

The somewhat different distribution of organizational sizes in the two studies is shown in table 4–1.

It can be seen that the Danielson and Danielson study incorporated a smaller proportion of organizations with more than 500 employees (about 7 percent), while the Fielding and Breslow study reported nearly 22 percent in this category. Table 4–2 shows that HPPs were more likely to appear in larger companies (>500) in the Danielson study. Fielding and Breslow's figures are not directly comparable, but they too show a correlation between the number of health promotion activities in an organization and the number of employees ($r = 0.43$).

It is interesting that both teams of investigators chose to include within their definition of HPPs what many students of the field would probably consider to be elements of EAPs, such as mental health counseling and drug and alcohol counseling. As noted elsewhere in this book, there are common features in the operation of both types of program; nevertheless, the people who run them tend to be quite different in terms of their professional backgrounds and orientation. In spite of the tremendous potential for synergy between EAPs and HPPs, it is doubtful whether practitioners in either field would appreciate being identified with the other at this point in history. Unfortunately, the "turf" battle is unlikely to be stilled by the indiscriminate lumping together of EAP and HPP components as though they were interchangeable. Every program type listed in table 4–2 is subject to tremendous variation in quality, scope, duration, and theoretical basis. Thus, the data given here can be read only as a rough guide to the prevalence of programs in the two areas considered. Yet, in spite of these reservations, it is striking that the two sets of figures are so close in so many cases.

In the most recent Ontario study, Dean, Reid, and Gzowski (1984), using a sample base similar to that drawn by the Danielsons, found that participation rates in fitness programs *declined* as size of organization increased. Thus, although 33 percent of workplaces with more than 500 employees re-

Table 4–1
Comparison of Two Health Promotion Studies According to Size of Organization

Study 1[a]			Study 2[b]		
# Employees	n	%	# Employees	n	%
<200	203	47.9	50–100	374	55.0
200–499	129	30.4	101–200	156	23.0
⩾500	92	21.7	201–500	102	15.0
			>500	48	7.0
Total	424		Total	680	

[a]Fielding and Breslow (1983).
[b]Danielson and Danielson (1980).

Table 4–2
Comparison of Study Results on Type and Prevalence of Health Promotion Programs

	Study 1[a]	Study 2[b]	
Type of HPP	Organizations with >100 Employees %	Organizations with >50 Employees %	Organizations with >500 Employees[c] %
Smoking Cessation	8.3	4.1	11.9
Weight Control	7.5	{ 6.4	{ 18.5
Nutrition	5.2		
Mental Health Counseling	18.4	10.4	29.7
Exercise and Fitness	11.6	13.0	36.0
Drug and Alcohol Counseling	18.6	13.6	39.0
Stress Management	13.0	5.3	15.2
	(n = 424)	(n = 680)	(n = 48)

[a]Fielding and Breslow (1983).
[b]Danielson and Danielson (1980).
[c]This column of figures is a subset of the column directly to the left of it, that is, the 48 organizations with more than 500 employees are a subset of the 680 organizations with more than 50 employees.

ported fitness programs, only 16 percent of employees in those sites were said to be involved, on the average. In contrast, 10 percent of organizations with between 50 and 100 employees claimed to have fitness programs, and 29 percent of their workers reported involvement. It is interesting, too, that between the time of the Danielson and Danielson study (1980) and the Dean, Reid, and Gzowski study (1984), companies had become less likely to foot the bill for these programs. In 1980, 45 percent of employers claimed that they subsidized fitness programs, while in 1984 only 31 percent did so. By 1984, the typical fitness program was sponsored or coordinated by the employer but managed by employees. They were led by paid outsiders (31 percent), paid staffers (28 percent), or by volunteers (27 percent). Fifteen percent of organizations with fitness programs had joint committees comprising representatives of employee associations and management.

In spite of problems with definition, it seems fairly safe to conclude from the foregoing studies that EAPs and HPPs sometimes appear under the same roof. In the following section, we look more closely at the characteristics of organizations in which this phenomenon occurs.

A Study of HPPs in the Context of EAPs

The data discussed in this section were collected in the course of a study, one purpose of which was to determine the prevalence of HPPs among a group of organizations known to have EAPs. A particular interest of the investigators was to discover from interviews with EAP and allied staff whether HPPs were seen as having any role in the prevention and early management of alcohol-related problems.

Method

In the initial phase of the study, most respondents were already known to at least one member of the research team, so that a phone call sufficed to introduce the survey ($n = 15$). Beyond this, we availed ourselves of lists obtained from local area EAP associations to whose members an introductory letter was sent under the flag of the association involved. In the case of the Metro EAP Association, we were blind to the identities of the subscribers, who were asked by mail to call the investigators if they wished to participate in the study. This resulted in ten successful contacts. A further thirteen interviews were carried out under the auspices of the Peel EAP Association. An additional seven contacts were made as a result of invitations from people who had heard about the survey or who were known to the principal investigator. In total, respondents from forty-five organizations were interviewed.

Because this was a haphazard rather than a random survey, we cannot declare to what population the results reported here can be generalized. On the other hand, the sample included both conservative and progressive organizations with regard to EAP so that we have a reasonable variation of practices, attitudes, and opinions to look at. In that regard, it provides some useful insights into the correlates of these different approaches including some indication of results, at least in terms of referral rates. It includes "leading edge" organizations and some that are quite traditional in both management style and approach to the future.

The study itself was conducted from March to August 1984, although thirty of the forty-five interviews were carried out in the last two months of that period.

The Respondents

In all, we spoke to fifty-eight representatives of the forty-five organizations. In every site, however, a primary respondent was identified. This was usually the person who held the greatest day-to-day responsibility for the EAP.

The principal parties involved were: twelve nurses; eight EAP coordinators; eight personnel managers; six medical directors; six human relations officers; and one union officer.

Types of Organization

Relative to their prevalence in Ontario, the industrial/manufacturing sector in our sample was well represented (our sample, 28.9 percent; Dun and Bradstreet, Ontario, 26 percent). However, finance, public administration, and government were over-represented (our sample, 48 percent; Dun and Bradstreet, Ontario, 25 percent). Correspondingly, transport, communications, retail and wholesale trade, and public utilities were under-represented in our sample (our sample, 24 percent; Dun and Bradstreet, Ontario, 49 percent).

Our sample was dominated by large organizations (>500 employees). This is not reflective of Ontario as a whole (our sample, 82 percent >500 employees; Ontario, 7 percent). Among those organizations reporting their figures, 28 percent employed a work force that comprised more than half females.

Characteristics of the EAPs

The mean age of the EAPs considered in this study was 5.8 years; with 31 percent being relatively new (2 years or less), 47.6 percent being of moderate age (3 to 9 years), and 21.4 percent being long-established (10 to 18 years). Over half the respondents (53.7 percent) reported that their EAPs had become broader in scope over the years, while 39 percent reported no such

Table 4–3
Problems the Program Is Meant to Deal With

	Yes		No		Under Other Program	
	n	%	*n*	%	*n*	%
Alcohol	45	100	—		—	
Drugs	45	100	—		—	
Cross Addiction	45	100	—		—	
Domestic	32	71.1	9	20.0	4	8.9
Mental Health	32	71.1	9	20.0	4	8.9
Legal	30	68.2	10	22.7	4	9.1
Financial	33	73.3	8	17.8	4	8.9
Housing	30	68.2	10	22.7	4	9.1
Daycare	31	70.5	9	20.5	4	9.1
Work Problems	31	68.9	10	22.2	4	8.9
Other[a]	25	64.1	10	25.6	4	10.3

[a]Other problems include physical health, retirement and career planning, leisure, education, anxiety, "life-style–related" problems, bereavement, and wife abuse.

changes. It is important to note that although all respondents or other representatives of their organizations had referred people for treatment at one time or another, there were three cases in which it was felt that no formal EAP existed. Similarly, ten respondents indicated that there was no written policy governing their EAP.

The majority of organizations in our sample offered broad-based programs dealing with everything from addictions to daycare problems (table 4–3). Sometimes, there were two programs; one addictions-specific, the other more general in scope. This appears to be a relatively infrequent practice.

Nearly half the respondents (48.7 percent) claimed that family members' problems could be dealt with under the EAP, while another 38 percent said they would refer family members to other sources of help. Only 13 percent said unequivocally that no help for family members was available.

The most common auspices under which EAPs in this study were run were personnel departments (44 percent) and medical departments (30 percent). In nine cases (21 percent), primary responsibility for the program was in the hands of an EAP coordinator or staff in a special EAP unit. In both cases, the key element was the perceived independence of EAP management from personnel and medical department management.

In two cases either a union or a joint labor–management committee was considered to be in charge of the program. In spite of these formal designations of responsibility it appears that nursing staff end up with day-to-day responsibility in a disproportionate number of cases, as evidenced by the percentage of nurses among our primary respondents.

Three-quarters of the companies involved in this study were unionized. Among these, 40 percent reportedly had active joint labor–management EAP committees while in the remainder the committees were in various stages of inactivity or had ceased to exist. The nonexistence of a joint committee, however, was not thought to necessarily imply lack of union involvement.

Referral Rates

Among the thirty-four organizations from which data were reliably available, the mean referral rate was reported at 4.8 percent for all problems combined. However, this figure is inflated by the presence of a few very high referral rates, the highest being 23 percent. This one figure relates to a situation in which all contacts with the EAP are listed as referrals even though they may simply have been telephone communications. If these "outrider" referral rates are removed from our calculations, the mean for all problems combined becomes 2.7 percent.

Among the referrals for all problems combined, it was estimated by nearly half of all the respondents who felt they could make an estimate that more than 36 percent of their cases were alcohol- or drug-related. However, about a quarter of the respondents saw 10 percent or fewer of their total cases as being alcohol- or drug-related, while the remaining quarter estimated 11 to 36 percent. The mean estimated proportion of such cases was nonetheless 47.71 percent.

Twenty-seven respondents provided their estimates of the proportion of alcohol- and drug-related referrals that were self-motivated. (There was no attempt to define the term "self-motivated" since in this study we were trying to do no more than broadly delineate the correlates of organizations in which major variations in orientation to EAP were developing.) Over 40 percent of all respondents who ventured an estimate said that over 50 percent of their alcohol- and drug-related cases were people who were self-motivated in seeking help for their problems. However, over a third of the respondents felt that this was the case in less than 5 percent of their cases, while about a quarter estimated between 6 and 50 percent to be self-motivated.

EAP Shortfalls in Identification and Referral

Forty of the forty-five principal respondents thought that their EAPs were not reaching some people who needed to be reached. Most of these people (85 percent) believed that it would be advantageous to the employer if the unreached group were to surface, but few were able to estimate how large this group might be. However, young single males, clerical workers, and senior managers, among others, were identified as groups least likely to be successfully penetrated by EAPs. Over 30 percent of respondents mentioned "stress" as a factor in the problems faced by the unreached group. Stress was considered to emanate from many sources and to manifest itself in many forms. For example, the type of work could be intrinsically pressured, as in marketing or policing. Time demands and lack of supervision coupled with low status were mentioned as stressors among clerical employees. The threat of layoff is an almost endemic stressor, which is rendered particularly acute in industries where new technology is being rapidly introduced. Women in the workplace, with the frequently multiple role demands placed upon them—particularly when they have younger children—were seen as subject often to dysfunctionally high levels of stress.

While dysfunctionally high levels of stress were considered to be problematic in and of themselves, respondents frequently made a connection between stress and excessive drinking. Of twenty-one respondents who held firm views on the subject, all but one agreed (eight of them vigorously) that excessive drinkers (as opposed to alcoholics) were a problem for their organizations. It was assumed that the same respondents believed alcoholics to be a problem.

When asked what types of intervention might be of value for excessive drinkers, however, only three suggestions emerged with anything like a constituency to support them. The first two suggestions involved the use of in-house stress management courses and healthy life styles courses. The former was advocated by 31 percent of respondents and the latter by 29 percent. They did not in most cases specify the form or content of such courses. The third major suggestion (27 percent advocating it) was really an amalgam of recommendations concerned with more effectively implementing the EAP that was already in place, rather than adding new components to it. For instance, although education of the work force about the philosophy, purpose, and procedures of the EAP is usually a part of the written or unwritten understanding about what the program will involve, it is a frequently neglected element, which can lead to atrophy. Similarly, training of managers and supervisors with regard to identification of problems, documentation, confrontation, referral, and follow-through needs constant attention. Elements such as career counseling, financial planning assistance, and retirement counseling are all meant to deal with specific sets of stressors and could be reinforced within existing EAPs without extending the already accepted concept of these programs.

Health Promotion Activities

Respondents were asked whether their organizations engaged in any health promotion activities, leaving the definition of this term to their own perceptions. Thirty-five respondents replied affirmatively. The activities to which they referred were as follows, rank ordered according to frequency of mention:

1. *Informational sessions* (usually on single occasions) through film and talks (n = 20; 44 percent). This included four addictions-specific seminars and one on nutrition.

2. *Stress management courses* (n = 14; 31 percent). Of these, four had been discontinued while two others were offered only to special subgroups within the work force. Thus, 22 percent is probably a more accurate figure for ongoing programs available to the whole work force. The range of content was very wide. The length of the courses varied between one and a half hours and two days.

3. *Smoking cessation courses* (n = 13; 28 percent). Again, these ranged from one-day to six-week courses.

4. *In-house fitness programs (practical)* (n = 12; 27 percent). Three such programs had been discontinued due to "lack of interest." The range was again enormous, from once-a-week "dancercise" to daily workouts with a full-time paid fitness coordinator.

5. *Subsidized fitness clubs* (n = 7; 15.5 percent). Subsidies ranged from 10 to 80 percent of the cost of joining community clubs or other organizations. (It is interesting to note that eleven respondents thought it would be worthwhile to subsidize memberships and the cost of taking courses.)

6. *Healthy life-styles courses* (n = 5; 11 percent). These were multi-session courses covering a variety of topics from nutrition and exercise to smoking, alcohol use, and stress management.

7. *Individual counseling* (n = 5; 11 percent). Some respondents implied that their HPP service extended to the provision of advice at the primary and secondary levels, that is, before problems had become so severe that agents of the employer felt the need to intervene.

8. *Weight loss courses* (n = 4; 8.9 percent). This included courses led by professionals external to the organization as well as by occupational (in-house) nurses.

Many organizations offered a number of the listed activities, so the categories do overlap. Other interventions mentioned were: hypertension screening (n = 3), back education (n = 3), CPR courses (n = 2), a money management course (n = 1), and a pregnancy planning seminar (n = 1).

The rate at which these health promotion programs and activities are offered is in some cases greater than reported by Danielson and Danielson in their 1980 random survey of Ontario workplaces even when only those organizations in their study with over 500 employees are compared with ours (see table 4–2). This holds for stress and smoking cessation courses. In the case of fitness and weight-loss programs, however, the Danielson and Danielson sample reported more activity.

Characteristics of Organizations with EAPs and HPPs

The proportion of organizations in which HPPs were reported was high so that detailed comparisons with those in which no such program was claimed to exist (ten cases) were rendered difficult. However, a few apparent trends emerged with such strength, despite the small number of non-HPP companies, that we feel it is worthwhile to report them, particularly since they point to lines of investigation for the future. The key points are as follows:

1. The industrial manufacturing sector was least likely to report HPPs (62 percent); the financial/public administration sector, the most (100 percent).

2. Organizations in our sample where less than a quarter of the work force was female were found to have less HPPs than organizations in which a greater proportion was female (67 versus 83 percent). It is conceivable that women place greater demands upon their employers for HPPs and that they are readier consumers of such services once provided. In chapter 5 we note that drop-outs from a healthy life-styles course at an industrial manufacturing plant were twice as likely to be men as women.

3. Broad-based EAPs were more likely to be found cheek by jowl with HPPs than were alcohol- and drug-specific EAPs; 91 percent of those EAPs that covered housing problems were found with HPPs, while only 40 percent of those EAPs not dealing with housing problems were found with HPPs. In addition, where a broad range of problems were covered by the EAP, the work force was likely to be over one-quarter female. For example, of organizations reporting up to a 25 percent female work force only two-thirds covered mental health; where more than 25 percent of the work force was female 92 percent of organizations covered the problem. The same statistics hold for problems associated with daycare for children.

4. In the thirty-four organizations whose representatives could provide us with statistics, a positive relationship emerged between the proportion of the work force that used the EAP and participation in an HPP. A full 100 percent of organizations reporting over 5 percent of the work force using the EAP also reported HPP participation, and 90 percent of those reporting 2 to 5 percent work force use of EAP also reported HPP activity. On the other hand, of those organizations reporting 1 percent or less work force use of EAP, only 57 percent also engaged in an HPP.

The proportion of women in the work force also appears to have a bearing upon the relationship between referral rates and health promotion. Where women represented more than one-quarter of the work force, the EAP was much more likely to be used by over 2 percent of employees. None of the organizations reporting that over 50 percent of the work force was female recorded EAP utilization at less than 2 percent, whereas 40 percent of organizations reporting less than one-quarter female work force so responded.

Where women comprised over half of the work force 50 percent of respondents reported that alcohol-related cases were less than 10 percent of all EAP cases. Where women comprised only up to one-quarter of the work force, however, 50 percent of respondents reported that more than 36 percent of their cases were alcohol-related.

From the evidence just reviewed we might conclude that the proportion of women in the work force is an important factor in the development and form of EAPs and HPPs.

5. When we look at alcohol and drug cases as a proportion of total EAP cases, a clear inverse relationship emerges between the percentage of substance-abuse cases and the likelihood of health promotion activity. Where organizations reported over 36 percent of all cases being alcohol or drug-related, 61 percent reported health promotion, while where alcohol and drug cases comprised 10 percent or less of total EAP cases, 100 percent of organizations were also engaged in health promotion, followed by organizations where alcohol and drug cases comprised 11 to 36 percent of total cases with a ninety percent involvement rate. Thus, the likelihood of health promotion programming decreased as the proportion of EAP substance-abuse cases increased. Conversely, the less emphasis an EAP placed upon alcohol or drugs, the more likely it was to be involved in Health Promotion. Caution must be exercised in the interpretation of this finding: recall that the *overall* EAP referral rate from organizations with HPPs tended to be much higher in this sample than among organizations without HPPs. Further, HPP employers tended to have a higher proportion of females on the payroll, and respondents felt that problems with alcohol decreased in proportion to the number of women in the work force relative to men. Thus, although alcohol-related referrals may have been less common in absolute terms in HPP companies, we have to consider their proportion to the overall caseload, which appears to be heavy. Further, there may be less need for alcohol-related referrals in companies with larger proportions of female employees. It would be presumptuous to even suggest at this stage of our understanding of HPPs that these programs have a preventive effect that to some extent forestalls the need for EAP referrals.

In spite of these cautionary notes, it is nonetheless possible that the low alcohol-related referral rates in HPP companies point to a problem that has been repeatedly predicted by some of the foremost scholars in the EAP field, among them Roman (1981b, 1984) and Trice (1983). The problem may be that the people who run broad-based EAPs and others who run HPPs either lose or never develop a sensitivity to the special difficulties associated with getting alcoholics into treatment. Consequently, the EAP may not emphasize the need for a constructive coercion component (Trice and Beyer 1984)—a component without which it is unlikely that "hard-core" alcoholics will be successfully managed or assisted.

Conclusion

We have shown in this chapter that HPPs are beginning to penetrate the workplace to a significant degree. While they vary enormously in nature and quality, they are clearly a force to be reckoned with simply by virtue of their presence. They coexist with EAPs in many cases. The ideological conflict, alluded to throughout this book, between EAP and HPP needs to be seen in light of this fact. Often, one gleans the impression that some EAP professionals feel they may opt in or out of dealing with health promotion professionals, and yet the frequent presence of both types of program in the same workplace suggests that representatives of each "side" should at least be seeking to establish diplomatic relations with the other. On the other hand, HPPs do not seem to be evenly distributed throughout the workplace: they are more common in the financial and public sector, and they are more likely to occur in the context of broad-based EAPs in worksites where a sizable proportion of employees are women. Currently, they are more common in larger organizations (>500 employees). Since the EAPs with which they are found tend to be highly active, HPPs seem to be part of an energetic approach to the well-being of all employees. Yet a warning must be sounded lest the needs of seriously troubled workers be ignored in an effort to go broad and deep with the health needs of the whole work force. Alcoholics, and possibly other people with florid mental health problems, may be particularly at risk while managers of the programs that were originally set up to assist them go on to greener pastures, forgetting the needs of the walking wounded. This need not happen, of course. Proactive management of EAPs and HPPs can surely avoid this state of affairs, but this requires a recognition from both camps that each has something to offer the other.

5

First-Stage Evaluation of a Health Promotion Course in an Industrial Setting: Impact upon Alcohol Use

with Harry Hodgson

This chapter describes the first evaluation of a prototype healthy life-styles program entitled "Take Charge: A Self-Help Course in Feeling Better and Living Longer." The purpose of this six-hour course was to provide employees at an automobile parts manufacturing plant with an opportunity to review their life styles, and within that context to consider in particular their cardiovascular health, their stress levels, and their use of alcohol.

The evaluation was based on before-and-after questionnaire responses from participating employees at the site where the course was conducted. The results presented here describe the changes in participants that strong circumstantial evidence suggests are attributable to the course, and so far lend support to the argument that among normal populations attention to general health-related beliefs, attitudes, and behaviors can bring about changes in immoderate alcohol consumption. The net impact of the course appears to have been positive; however, some evidence of negative effects was also found.

Origin of "Take Charge"

The designer of the course, Harry Hodgson, is a community consultant with the Addiction Research Foundation of Ontario. Hereafter, he will be referred to as the course convener. "Take Charge" grew out of an experimental course conducted for the work force of a public utilities company in Windsor, Ontario. This earlier course consisted of three major educational components related to healthy heart maintenance, moderate alcohol consumption, and stress management. At that time, the educational package reflected the stated needs of the work force according to its representatives on the health and safety committee of the client company. The prototype program was very well received according to participant ratings. The course convener was sufficiently encouraged by this response to invite the first author of this book (hereafter called the investigator) to design an evaluation for the program that was to be repeated in another company, namely the automotive parts manufacturing organization that is the subject of this report.

An agreement was reached between the convener and the investigator to carry out an evaluation of a modified version of the course. The modifications were concerned primarily with focusing the objectives of the program and with developing greater thematic continuity. The suggested modifications were consistent with major theoretical propositions and empirical evidence concerning the manner in which health-related beliefs, attitudes, and behaviors can be effectively changed. Thus, the investigator's initial task was simply to compare the practical and theoretical models, suggesting closer links where warranted. A guiding principle in this preliminary work was that the links between a general health promotion course (as requested by the client company) and the prevention of alcohol abuse (the mandate of the convener and investigator, both of whom are employed by the Addiction Research Foundation) should be made clear in the objectives of the course and in the evaluation design.

Purposes of "Take Charge"

The reformulated purpose of conducting the course was to examine its potential usefulness in (a) reinforcing the life-style supports for moderate (low-risk) consumption of alcohol, and (b) reducing high-risk consumption. The course was thereby designed to incorporate primary- and secondary-prevention objectives. A minor tertiary-prevention objective was to encourage those who were in need of clinical intervention to seek help.

The conceptual basis of the course was that alcohol-related behavior is supported by a complex system of more general health-related beliefs, attitudes, values, and behavior. In order to effect lasting change in alcohol-related behavior it was argued that this supporting system would have to be modified also. Consequently, the course was designed to:

1. Influence beliefs and attitudes about health and how to improve, maintain, or restore it.
2. Promote awareness of alcohol use in the context of life style as a system of interrelated behaviors.
3. Stimulate the formation of intentions to make changes in alcohol use and in other behaviors that may encourage or support its excessive use.
4. Promote actual changes in the areas of alcohol use and other behaviors that may encourage or support its excessive use.
5. Stimulate high-risk alcohol users to seek help.

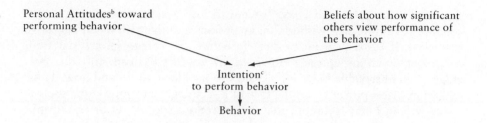

Personal Attitudes[b] toward performing behavior

Beliefs about how significant others view performance of the behavior

Intention[c] to perform behavior

Behavior

[a]Adapted from: Fishbein M., *Factors influencing health behaviours: an analysis based on a theory of reasoned action.* In F. Landry (ed.) Health Risk Estimation, Risk Reduction and Health Promotion. Papers presented at the 18th Annual Meeting of the Society of Prospective Medicine, Quebec City, Oct. 20–23, 1982. Canadian Public Health Assn., Ottawa, Canada 1983.

[b]Personal attitudes are seen as deriving from *beliefs* about the *consequences* of performing the behavior and from *evaluation* of these consequences.

[c]Intentions are seen as predicting behavior to the extent that i) the behavior is stated in specific terms as a discrete action, and ii) the number of steps between intention and behavior is small and the steps are capable of quick execution.

Figure 5–1. Simplified Fishbein[a] Model of Influences upon Health-Related Behavior

Theoretical Model of Change

Fishbein (1983) has incorporated a number of theoretical threads into a conceptual fabric that he describes as a "theory of reasoned action." In its entirety, this is a complex model. We present only a simplified version of it here, since the evaluation of "Take Charge" is not intended to be a test of any particular theoretical model; nonetheless, it is instructive to locate the course within the theoretical context of Fishbein's model since it is perhaps the most elegant exposition of the variables that are currently thought to affect health behavior in a direct manner. Figure 5–1 presents the main elements of this model.

Note that there is no mention of socioeconomic and demographic influences. These are considered to be potent but beyond the influence of most educational programs. Also, they are "givens" in any model of change (Fishbein 1983).

In Take Charge, the aims were to affect those beliefs that were assumed to influence the participants' intentions to change their health behavior and to suggest ways in which intentions could be realized in action through the designation and execution of small, self-reinforcing steps toward clear, simple behavioral objectives. As noted in figure 5–1, the likelihood that an intention will be realized in action is seen as being a function of how specific the behavioral objective is, and how clear and few the steps are between thought and deed. For example, if I say, "I want to lose weight" (a behavioral outcome), I am confronted with too many options and the path to my objective is unclear. If I say, "I want to diet" (a behavioral category), at least I am narrowing down my options. However, if I say, "I will drink only diet beverages" (an observable behavior), I know what I have to do and how. Thus, while I may feel powerless in the face of a seemingly enormous task such as losing weight, I may feel more able to undertake a specific set of acts (drinking diet drinks) because they are simple, feasible, and bear a clear relationship to my ultimate objective (losing weight).

The concept of powerlessness in the example just given is extremely important. The rationale for developing highly specific behavioral objectives (small self-reinforcing steps toward some valued state or condition) is strongly tied to the sense of powerfulness or *self-efficacy* that is generated from the successful execution of such plans (Bandura 1977a). Of course, the sense of self-efficacy becomes relevant only after some need to act has been established. In health terms, a threat must be demonstrated before it can be expected that the behavior producing the threat will be changed. Accordingly, Beck (1983) argues that health messages designed to change people's health behavior must persuade the target audience that a real threat to their well-being exists and that they can control this threat through their own actions. In a study with Lund, Beck adds that the sense of control that must be engendered with regard to the threat is probably best developed in precise behavioral terms. That is, personal efficacy is best understood and best encouraged in relation to specific behavioral objectives rather than in global terms (Beck and Lund 1981). This echoes Fishbein's hypothesis that intentions best predict behavior when the latter is described in highly specific terms.

Course Content

The gist of the Take Charge program was that all employees who wanted to attend were asked in the course of three 2-hour sessions to evaluate their overall life styles and the role of alcohol within them. They were exposed, during work hours, to printed, filmed and spoken messages about practices that are believed to contribute to or to take away from the risk of suffering from certain conditions such as cardiovascular diseases, some forms of cancer, pulmonary disease, and so on. Alcohol abuse and stress were portrayed as risk factors for premature morbidity, mortality, and deficits in the quality of life. The interactivity of life style practices was explained in such a way as to illustrate the manner in which alcohol use becomes woven into the fabric of everyday life. The components of life style—how we eat, drink, sleep, work, recreate, and manage stress—were described as parts of an interdependent system of practices, attitudes, beliefs, and values. The concept of personal control over the development of this system was explored, and participants were asked to identify areas of their lives that could be controlled under their own direction and at their own pace. Suggestions were made concerning strategies that could be used in making changes, although the course was designed primarily to influence participants' awareness and acceptance of the need to change rather than to bring about lasting behavioral changes, even though signs of the latter were sought. We believed that behavioral change of a lasting nature required more attention to skill building and reinforcement than was given in this course: our intention was to find a language and a means by which to communicate with participants about life style as a system and the role of alcohol consumption within it. (See appendix for an outline of the course.)

Method of Enquiry

The conduct of the course and its pre- and post-evaluation in a single site, with no external comparison or control, was considered to be an exploratory undertaking. Consequently, the results reported here, hopefully of great interest to those involved in the practice of and research on alcohol-abuse prevention, should be seen as a contribution to the future design of more definitive studies on the effectiveness of programs.

Risk group analysis is the term that we employ to describe our first approach to analysis of the data. It called for the categorization of the sample into four groups defined according to their risk of being or becoming excessive drinkers. This form of analysis reflected the varying objectives of the course, which were themselves stated in relation to risk levels. Thus, the reinforcement of life style supports for moderate drinking is an objective that possesses meaning largely in reference to a group whose current drinking practices can be considered moderate. Similarly, the reduction of consumption levels among high-risk drinkers clearly has meaning only in reference to one group. The analysis in the first part of this chapter largely follows the risk group format.

Factor analysis was subsequently used to investigate the structure of relationships among the measures of beliefs, attitudes, and behaviors before and six months after the course. Changes in this structure are helpful indicators of the ways in which an intervention such as Take Charge affects participants. However, these changes are rarely investigated in studies of health promotion impact, with the result that certain types of effect are not noticed. In particular, we think that small changes in a number of related areas are likely to be missed by univariate and bivariate analyses when in fact the overall shift is of practical significance. These changes may not be important in themselves, but when considered as part of a correlated set they may become significant. Factor analysis allows us to inspect changes of this sort. In addition, factor analysis, by not assuming that one attitude or behavior is dependent on another (as in multiple regression), allows us to see the full interdependence within the total data set. The results of this analysis are presented in the second part of the chapter.

Questionnaires were administered to course participants at two points: once before the first session of the course, and again approximately six months later.

Health-Related Measures

Various measures were developed or adapted for the evaluation, most of them in some way relating to the simplified Fishbein model of change shown in figure 5–1. These measures include the following.

1. *Values placed on personal/family health.* These values were incorporated into an original list of values compiled by Rokeach (1973). Respondents were asked to identify five values that were of greatest importance to them. Our purpose was to determine whether health was placed among the top five values, and whether the course brought about any shift in these priorities. This variable was introduced in order to determine the extent to which health behavior is influenced by evaluation of health relative to other goals, states, and conditions.

2. *Importance of health-related behaviors and conditions.* Respondents were asked to rate each of six health behaviors or conditions according to their importance in their own lives, for instance, "Knowing how to protect my physical and mental health"; "Feeling well physically/mentally"; "Eating foods that are good for me"; "Getting enough sleep/exercise." Again, these categories were designed to examine the relationship between the salience of health and health-related behaviors.

3. *Locus of health control.* Six items were developed in order to determine beliefs pertaining to the degree of personal control that respondents perceived over their own health, for example, "My personal life style is something that can affect my future health"; "I can control the effect that the stresses of life have on me." These items were scored to produce a single measure of perceived health control. We expected that this variable would be correlated with the general measure of self-efficacy (measure 4, "mastery"); however, consistent with doubts raised by Beck and Lund (1981) and McAlister (1983), it appears from our data that self-efficacy is best defined relative to specific areas of competence. Thus, one may feel very much in control generally, but quite lacking in specific areas such as health.

4. *Mastery.* Pearlin and Schooler (1978) developed this seven-item scale to measure general sense of control or self-efficacy, for example, "I have little control over the things that happen to me"; "What happens to me in the

future mostly depends on me." It was anticipated that the willingness to form intentions to change health behavior would be positively related to high scores on both mastery and locus of health control.

5. *Reported health problems in self and family.* A score was developed from five questions relating to illness and debilitation in self and family during the last year. While not directly related to the Fishbein model, it was anticipated that the experience of illness, whether personal or familial, might influence health behavior through modification of personal attitudes toward preventive action.

6. *Factors believed to contribute to excessive drinking.* Respondents were asked to indicate how likely they considered nine factors to be in contributing to excessive drinking. Six of the most frequently cited factors were used to generate an overall score. These were: "Having too much stress in your life";"Irregular working hours (for example, overtime, bringing work home, working weekends, shift work)"; "Dull or frustrating work"; "Smoking"; "Being unhappy or frustrated in some area of your life"; "Feeling uncomfortable with people." All these factors were mentioned during the course as being potentially linked, alone and in concert, with the development of excessive drinking.

7. *Signs of problem drinking recognized.* The signs were those referred to in the film "Early Warning Signs" (see appendix, session 3). Respondents indicated the perceived likelihood that eleven behaviors were signs of a developing drinking problem, for example, "Ability to drink increasing amounts and not get drunk"; "Drinking alone"; "Occasionally getting the shakes." Responses to these items were combined to produce a score.

8. *Intentions to change.* Participants were asked whether they intended to make any changes for themselves or on behalf of their families in the areas of eating, exercise, leisure time, alcohol use, smoking, work, and sleep. These intended changes were considered individually and collectively. (*Actual reported changes* were monitored in the areas listed under "Intentions.")

9. *Trait anxiety.* The Spielberger Trait Anxiety Scale was administered at baseline (pre-course) only (Spielberger, Gorsuch, and Lushene 1970). The original purpose of collecting data in this area was to examine the relationship between trait anxiety (a general tendency to react to threatening situations in an anxious manner) and alcohol use. We wished to test the general hypothesis that excessive alcohol use is related to abnormally high levels of anxiety among normal as opposed to clinical populations. By using the Tension Reduction Drinking Scale (Parker and Brody 1982) we wished also to test the hypothesis that people who drink excessively do so in order to relieve their anxiety or to otherwise alter their mood. These hypotheses are not related directly to the Fishbein model, although they have an oblique bearing upon it in so far as beliefs about personal efficacy are involved in the tension reduction hypothesis of drinking (Cappell 1975). This hypothesis asserts (1)

that alcohol reduces tension physiologically, but also (2) that alcohol is used with the expectation (belief) that it will reduce tension. As noted in chapter 6, it is possible to confirm the latter without having proved the former, that is, there is a belief/expectation effect associated with the use of alcohol that is different from the actual physiological effect.

10. *Factors believed to contribute to cardiovascular disease.* The sessions of the course on cardiovascular health (appendix, sessions 1 and 2) provided participants with current information on risk factors associated with cardiovascular disease, namely: family history of heart disease; smoking; hypertension; hyperlipidemia (resulting from too much fat in the form of cholesterol and triglycerides in the diet); stress; obesity; and lack of sufficient exercise. Respondents were asked to ascribe levels of importance to these factors before the course and at follow-up. While these items appear to be of a purely informational nature, it must be appreciated that their acceptance as factual depends a great deal upon the pre-existing beliefs of the audience and upon the credibility of the information source. To some extent, we expected that the degree of importance ascribed to these cardiovascular risk factors would be related to other "risky" attitudes, beliefs, and practices. The items were treated collectively as a score for purposes of analysis.

11. *Demographic and socioeconomic indicators.* Age, sex, marital status, number of children, education, and occupation were introduced as potentially discriminating variables with regard to health beliefs, behavior, and differential program impact. No specific hypotheses were advanced. However, since our target population was in large part blue-collar we were keenly interested in the reception that would be extended to the course since so much of the experience with health promotion programs had hitherto been gathered in relation to middle-class, white-collar groups.

12. *Social support.* In the Fishbein model the individual's beliefs about how significant others view his or her behavior are seen as important influences upon the formation of intentions to change. Indeed, the support of others is an important factor in the success of many a health venture, be it weight loss, smoking cessation, getting more exercise, relaxing more, obtaining more sleep, drinking less, and so on. Consequently, we were interested in exploring whether differing levels of support affected the probability of participants' reporting the successful adoption of new health-related behaviors in any of the areas just noted. Respondents were asked to record support or the lack of it from all likely sources, according to the areas in which they wanted to make changes.

13. *Barriers to change.* There are numerous barriers to making health-related changes, both real and imagined. We enquired about perceived constraints of time, money, lack of support, know-how, and will power. In the Fishbein model, barriers are presumably perceived less often when the

changes that are attempted are small and manageable, so that the gap between thought and deed (intention and behavior) is as small as possible.

14. *Drinking behavior and beliefs.* Actual consumption was measured according to frequency and amount. Definitions of "social drinking" and the "safe" number of drinks that can be consumed in one hour before driving were also elicited from respondents. Actual consumption, controlling for body weight, formed the basis for constructing "risk groups" defined in relation to the hazards represented by increasing levels of alcohol intake.

Results

In this section we describe the participants and the two major approaches that were used to interpret the data from the "before and after" evaluation—risk group analysis and factor analysis.

Attendance

The course organizers (the convener and representatives of the company) were able to recruit and hold a sizable proportion of the work force throughout the six sessions of the course and were able to persuade them to return for the follow-up six months later. This was a most encouraging fact in itself, since we were afraid that we might end up preaching to the converted, a problem that has tended to characterize health promotion programs of this type. As figure 5–2 illustrates, nearly 74 percent of those eligible to take the course (twenty-four people were excluded because they had been involved in a pilot version of Take Charge) appeared at the first session. Between the first and last sessions there was less than a 25 percent drop-out rate. Thus, over 50 percent of the work force completed the course in its entirety. Of this proportion, almost all were involved in the evaluation. Among those who completed both questionnaires, sixty-five were males (54.2 percent) and fifty-five were women (45.8 percent).

Research Completers vs. Research Drop-Outs

A question that frequently occurs in evaluation research is whether those participants who completed the research instruments on a pre- and post-basis were different from those who completed only the first administration. In our case, most of the 30 percent drop-out took place between the first and last sessions of the course (figure 5–2), very little occurring between the last session and research follow-up six months later. It appears that drop-outs did not differ significantly from completers in terms of major health character-

Dates of Sessions	Course Attendance		Research Compliance
	On Payroll and Eligible	Course Attendance	Questionnaires Completed
Oct. 1982 (Pre-test) (1st Session)	294 (another 24 had taken course earlier)	(session 1) 217 ← (73.8%)	206 ← (94.9%)
Nov. 1982 (Last Session)	294	(session 6) 165[a] ← (56.1%)	No questionnaires administered
April 1983 (Post-test Only) (No Session)	280	152[b] ← (54.2%)	146[c] ← (96.9%)

Notes

[a]Drop-out between sessions 1 and 6 was 24 percent.

[b]Drop-out between session 1 and research follow-up was 30 percent.

[c]Initial match rate of questionnaires, pre-course to follow-up was 71 percent. However, only 126 of the 146 post-tests were matchable according to the self-generated codes used to link pre- and post-questionnaires.

Figure 5–2. Pre- and Post-response Rates of Company Employees

istics. However, they appear to have been somewhat better educated, 20 percent having gone beyond high school in comparison with 11 percent in the sample as a whole and to have been more likely to hold white-collar positions (24 percent of drop-outs so reporting versus 3.4 percent among completers). Similarly, another 18 percent of the drop-outs were skilled tradespeople in distinction from the completers among whom only 4.3 percent were so classified. Given that there was no coercion to attend the course, it is interesting that blue-collar workers chose to attend more than white-collar workers. Other explanations for this occurrence exist, however, beyond those associated with the appeal of the course. Since Take Charge was run during work hours (the first hour of each shift on one day a week for six weeks), it was necessary to shut down the assembly line in order for any employees involved in that process to attend the sessions. Thus, if the line was shut down, the only alternatives for line workers were attendance at the course or socializing with others who did not want to attend. Everyone had to be in the plant, however, so that it was not an option to come in late. White-collar employees, on the other hand, had the choice of returning to work if they wished, since their jobs were not dependent on those of others in the same way that assembly-line workers' jobs were.

A further characteristic of drop-outs was that they were much more likely to be male than female (47 percent of the males in attendance at session 1 dropped out by follow-up, compared with 27 percent of the females).

Health Characteristics Related to Age, Sex, and Education

The following significant or nearly significant differences were noted among research completers.

Age and Health Characteristics (Baseline). A difficulty with determining reliable trends as related to age was that few people fell into the very young or old categories relative to the mean age of the sample as a whole (42.3 years). Thus, only six people were 29 or younger and only five were 60 or older. Given this limitation, although younger people tend not to report greater concern about health as measured by the importance of health behavior score, they do report a greater readiness to make health-related changes as measured by the sum of intended changes score ($X^2 = 13.36664$; $P > .10$). If we take the lowest category of summed intended changes we find that with age there is a linear increase in unwillingness to make changes. Among those under the age of 30 only 17 percent score in the lowest category compared with 80 percent among those over the age of 60.

However, there is a departure from the linearity of this trend among the 40–49-year-old group in that although there is a large proportion in the lowest category of intended change, this group also contains the largest proportion of people in the highest category (41.4 percent vs. the expected 29.6 percent for the sample as a whole).

Members of the 40–49-year-old group are remarkable also with regard to their low scores on the Trait Anxiety Scale (Spielberger, Gorsuch, and Lushene 1970). Only 21 percent score above 40 compared with the 38 percent expected in a normal population. Conversely, 57 percent of those in the 50–59-year-old group score above 40, thus manifesting the *highest* rate of such scores on the Spielberger Trait Anxiety Scale ($X^2 = 19.51903$; $P > .08$). While results of this nature are difficult to interpret in themselves, it is worth speculating that the life phases of participants have a bearing upon the degree of anxiety that they experience. The time during which children have left home and begun to make their own way in life (when parents are in their 40s) may be a time of relative tranquility compared with the anxiety of pre-retirement years (the 50s). In terms of audience segmentation with regard to different needs, this kind of preliminary observation deserves greater attention from researchers and programmers.

Age and self-reported absence from work are related at the age extremes, with the youngest group reporting the lowest rate of absence for five days or more in total (0 percent) and the oldest group reporting the highest rate (75 percent). However, the trend in between these extremes is not linear, since the 50–59-year-olds have a strikingly low absence rate (15 percent).

The same trend prevails when absences for more than three consecutive days are examined. The combination of high anxiety and infrequent absence from work suggests a concern about job security among the 50–59-year-old group or perhaps a more generalized concern about retirement which is possibly resolved by the early 60s. We hasten to repeat that this is a highly speculative interpretation.

The expense of making health-related changes tended to be perceived as a barrier to change by younger respondents. Thirty percent of those between the ages of 30 and 39 reported this problem compared with 15 percent in the 40–49-year-old group and 4 percent in the 50–59-year-old group ($X^2 = 8.81107$; $P > .07$). One might speculate that lower incomes associated with junior positions and the financial burden of child rearing account for some of these differences.

While it was not a statistically significant finding, it is worth noting that "lack of know-how" as a barrier to change was reported more frequently by the 40–49-year-old group than by any other (37 vs. 26 percent for the whole sample).

Sex and Health Characteristics (Baseline). Women in this sample appear to be more interested than men in making health-related changes, judging by the sum of intended changes score. Seventy-four percent of the females, compared with 55 percent of the males, said that they wanted to make more than a few changes. Lack of time as a barrier to making changes was more important for women than for men (F: 41 percent, M: 33 percent), while lack of money was more likely to be perceived as a barrier by men (M: 22 percent, F: 12 percent). Individually, these differences were not statistically significant, but taken together they may have some practical significance. Women in our society tend to be the incumbents of more roles than men, often having actual or perceived principal responsibilities for child rearing and homemaking even when employed outside the home on a full-time basis. This added burden may be reflected in sex differences found in trait anxiety where women score higher than men. Using the ≥ 40 raw score criterion used in the foregoing section, we find that 45 percent of the females in this sample scored in this high range compared with 30 percent of the males ($X^2 = 7.19187$; $P > .07$).

Females also reported more days of absence than males. Among women, 29 percent reported 5 days or more total absence during the preceding year compared to 19 percent among men. Whether this is a reflection of personal health problems or of responsibilities for family members or of some other factor cannot be answered through the data at hand. However, in a pilot study of a healthy life-styles course for women (Pupo, Shain, and Boutilier 1985), it was found that among those with children at home 50 percent reported absences of more than three consecutive days compared with 22 percent of those without children at home.

Education and Health Characteristics. Using the completion of high school as the criterion, we find that at baseline, people with less education (high school or less) were more likely to place family health among their top five values according to the modified Rokeach list (59 percent of this group versus 29 percent of those with more education. $X^2 = 11.20581$; $P > .03$).

This trend is echoed in the importance of health behavior where 71 percent of the less educated, versus 57 percent of the more educated people were placed in the higher scoring category ($X^2 = 16.38376$; $P > .04$).

However, an apparent contradiction occurs in the expression of intended health-related changes: people with more education were somewhat more likely to express intentions to change than others ($\leq$ high school: 28 percent; $>$ high school: 43 percent reported intentions to change in more than six areas). This was not a statistically significant difference, but it may reflect a differential perception related to educational background as to how accessible or feasible such changes are.

Similarly, the salience of family health as a value among less-educated people in this sample may be a reflection of greater concern among those with less perceived personal resources for the betterment of their family's lot. This interpretation, however, requires more fundamental testing than is possible with the present data. It is necessary, at least, to control for the influence of family responsibilities (for example, children or elderly parents in one's care) when considering differences in the value placed upon family health.

Conclusions on Health Characteristics Related to Age, Sex, and Education

There appear to be some important differences in health outlook and behavior depending on age, sex, and education. These differences may indicate a need for the formulation of health messages designed for specific subgroups defined according to at least demographic characteristics. This need not involve physical segmentation of audiences. In some cases, it could mean simply the inclusion of messages of particular significance to subgroups within an audience.

Risk Group Analysis

As noted previously, four "drinking risk" groups were constructed. They were built by dividing body weight in pounds by alcohol in standard drinks per day and applying the resultant figures to the Risk-O-Graph (Addiction Research Foundation, n.d.). The Risk-O-Graph plots body weight against the mean reported number of drinks per day to yield estimations of risk based on any two values of the variables. Risk levels are broadly classified as "safe, increased risk (which we will also refer to as cautionary), and dangerous." According to this manner of conceptualizing the risk associated with excessive drinking, a person of medium muscular build weighing 170 pounds consuming four standard drinks per day on the average is on the borderline between safe and cautionary levels of consumption. A person weighing less than 170 pounds, let us say 130 pounds, is squarely in the cautionary range, while a person of 100 pounds would be drinking at hazardous levels. Conversely, a person of 220 pounds—assuming the same muscular structure—would be relatively safe according to these estimates (Addiction Research Foundation, n.d.).

The division of weight by consumption yields metric values that remain constant if the ratio of the two variables remains constant. Variations in the ratio serve to indicate higher or lower risk, so respondents can be assigned scores that reflect this. Cut-off points can then be designated that place respondents in groups. In this case four groups were designated. The drinking-related characteristics of these risk groups are shown in table 5–1.

The female groups were based on risk score cutting points that were lower than those of males since, even controlling for body weight, there were fewer heavy drinkers among females. Consequently, as table 5–1 shows, risk group 4 for women is characterized by a lower weekly average consumption of alcohol (18.3 standard drinks) than among men, where the corresponding figure is 48.0 drinks per week on the average. Among women, therefore, the fourth risk group (n = 6 or 10.9 percent) comprised people who drank somewhat in excess of 2.6 drinks per day on the average but who were thereby placed in the "increased risk" or "cautionary" category rather than in the hazardous category according to the Risk-O-Graph criterion. Among men, however, the fourth risk group contained people who drank at the hazardous level (n = 5 or 7.7 percent). In spite of the difference in cutting points for men and women, the graduation from group 1 to group 4 nonetheless represents a continuum of risk.

Since the number of drinks per week is part of the definition of the risk groups, it is to be expected that they can be best differentiated on this basis. The groups varied correspondingly in their definitions of social drinking and the safe level of drinking before driving; drinks consumed per typical drinking occasion, and their mean scores on "tension reduction drinking." These differences were confirmed as statistically significant ($P > .01$) by one way analysis of variance.

A major purpose in constructing risk groups in this way was to provide a basis both for examination of the health-related correlates of drinking behavior and upon which differential effects of the course relative to drinking status could be observed. Before proceeding to the results of these analyses, we present those variables of a demographic nature that differentiated the

Table 5–1

Baseline Drinking-Related Characteristics of Four Risk Groups Defined by Weekly Alcohol Consumption, Controlling for Body Weight

	Low ← Risk Groups: Mean Values → High							
	1		2		3		4	
	M	F	M	F	M	F	M	F
n =	17	19	34	22	9	8	5	6
# of Drinks, on Average, per Week[a]	0	0	7.0	4.3	19.7	8.4	48.0	18.3
# of Drinks per Typical Drinking Occasion[b]	1.4	1.2	5.7	5.2	6.9	4.4	7.0	8.7
Definition of "Social Drinker" according to # Drinks per Week"[c]	3.6	3.4	10.4	5.7	20.3	9.4	35.0	18.7
Definition of "Safe" # of Drinks in One Hour before Driving[d]	1.0	0.9	2.7	2.3	2.9	2.1	2.4	2.0
"Tension Reduction Drinking" Score[e]	4.1	4.4	6.7	6.3	7.6	9.3	9.8	9.3

[a]Number of standard drinks per week. This is the sum of beer, spirits, wine, and fortified wine where the equivalence is given as follows: 12 oz. bottle of beer ≃ 1½ oz. spirits ≃ 3½ oz. fortified wine ≃ 5 oz. glass of wine.

[b]"How many drinks do you consume on a typical drinking occasion, for example, with friends on a Saturday night?" (uses same equivalence as above).

[c]"How many drinks per week, on the average, do you think you could consume and still be a moderate (social) drinker?"

[d]"How many drinks do you think you could consume during one hour and still be able to drive safely?"

[e]Derived from responses to the following items: "I drink to cheer myself up; I drink when I am tense and nervous; I drink to help me forget my worries; I drink to change the way I feel." Response range: "Often, occasionally, rarely, never." (Parker and Brody 1982)

risk groups. Similarly, certain variables that were measured only before the course are discussed here.

Age. Risk group 4—the highest risk group—was significantly older among male completers but not among female completers (table 5–2).

Marital Status. A higher proportion of men than of women were married (85 vs. 45 percent). While risk status among men did not vary significantly according to marital status, among women a higher proportion of the highest risk group were living together (44 percent vs. mean of 8 percent for the whole female sample), as opposed to being married.

Children. There was a slight but not significant tendency for males in

Table 5–2
Age and Risk Groups

	Low ← Risk Groups → High							
	1		2		3		4	
	M	F	M	F	M	F	M	F
n =	17	19	34	22	9	8	5	6
Mean Age	43.1	44.1	40.1	43.9	38.4	37.0	56.2	40.2

ANOVA (one-way): Males, F = 4.784, P > .005; Females, F = 0.968, P > .42.

high-risk groups to have children who were either older than seven years old or who had already left home.

Education. There was a tendency among both males and females in the highest risk group to be less formally educated than the others. This was more obvious with males, since 100 percent of that group had not completed high school, compared to the rate for the sample as a whole, which was 57.5 percent.

Reported Health Problems in Self and Family (Baseline Only). Males in the highest risk group were least likely to report such problems (33 percent reported no such problems against the average of 14 percent for the whole sample). Females in the highest risk group, however, were *most* likely to report problems of this kind. Zero percent in this group reported no problems compared with 15 percent in the sample as a whole. The lowest risk group (people who drink hardly at all or never) among females also reported a high incidence of health problems either pertaining to themselves or to their families.

An important consideration to bear in mind is that high-risk female drinkers (group 4) consumed less than their male counterparts (2.6 drinks per day on the average in comparison to nearly 7 drinks per day respectively). Consequently, some of the apparent sex differences are attributable to this relative definition of risk. For example, we find that males in risk group 3 (moderate–high drinkers) report the highest rate of physical health problems. Risk group 3 for males is equivalent to risk group 4 for females in terms of consumption (controlling for body weight). Consequently, the pattern for the sexes may not be different. We cannot say with certainty because of the under-representation of very heavy drinkers among the women in this sample. The tendency for men to report no health problems in the highest risk category arouses suspicions about denial of any difficulties (possibly perceived to be drinking-related) on the part of this group.

A similar suspicion is aroused when we examine self-reported absence from work during the last year according to risk group. Among men, we find that only 8 percent of the two highest risk groups report absences of more than three consecutive days compared with 22 percent of the two lowest risk groups. Among women, the figures are less dramatic: 17 and 25 percent respectively.

While these findings may raise suspicion about the validity of self-reported absence, they may also reflect a certain protectiveness on the part of heavier drinkers who wish to avoid confrontation about their drinking by keeping up regular attendance. Alternatively, the figures may not be hiding anything; perhaps they reflect neither denial nor protectiveness, simply a robust constitution. However, one should bear in mind that within the heavy drinkers group is a subgroup of excessive drinkers. We know from other accounts that such people are absent a great deal more than their more moderate colleagues.

Trait Anxiety (Baseline Only). Among males there were no significant differences between risk groups. Among females, however, a near-significant trend emerged (see table 5–3). As consumption of alcohol increases, trait anxiety appears to *decrease*, although the trend is not consistently linear. However, it is also important to note that among low-risk respondents, female trait anxiety was higher than among males. In fact, female trait anxiety scores among the low-risk group fell squarely within a range of such scores reported among women participating in a stress management course (Herbert and Gutman 1980). It is interesting to recall that "tension reduction drinking" scores for both women and men increase as consumption increases (see table 5–1). Thus, in the case of females, we appear, on the surface, to have a situation at baseline in which tension reduction is actually accomplished through the use of alcohol, or where at least drinking is part of a seemingly successful coping strategy, since among female high-risk respondents, trait anxiety scores are well within normal bounds. Again, however, it is important to bear in mind that high-risk among females in this sample is a relative term.

Male scores on trait anxiety are not remarkable, falling well within normal bounds as established by the test originators (Spielberger, Gorsuch, and Lushene 1970).

Trait anxiety was found to correlate significantly with mastery ($P > .001$) among both males and females (-0.66 and -0.63 respectively). The negative relationship indicates that as anxiety increases, mastery (self-efficacy) tends to decrease (see chapter 6). It is worth noting, however, that it is far from obvious that this is a one-way relationship. Conceivably, a poor sense of mastery could lead to high anxiety just as easily as the reverse.

Profile of the High-Risk Drinker at Baseline. The typical high-risk drinker, then, was someone who, before the course: consumed the equivalent of more than four standard drinks per day, if male, and nearly three if female; consid-

Table 5–3
Trait Anxiety[a] Scores and Risk Groups

			Low ←	*Risk Groups*	→	*High*		
	1		2		3		4	
	M	F	M	F	M	F	M	F
n =	17	19	34	22	9	8	5	6
Mean Scores	36.0	43.3	35.8	36.9	37.1	41.8	35.8	36.2

ANOVA (one-way):   Males, F = 0.052, P > .98; Females, F = 2.583, P > .06.
[a]Spielberger, Gorsuch, and Lushene (1970).

ered themselves social drinkers; felt they could drive after more than two drinks in the previous hour; and tended to drink to relieve tension or otherwise change mood. Among males but not females, higher risk was associated with greater age and the likelihood of having older children some of whom had already left home. Higher risk males denied having health problems or higher than average absenteeism. Females in the higher risk groups, on the other hand, reported more health problems than others but they too reported no higher than average absenteeism. Both men and women in the high-risk categories tended to be less educated than their lower risk counterparts. Keeping in mind that high-risk for women in this sample is a relative term, trait anxiety tended to be more elevated among low-risk than among high-risk females. Among both men and women, a significant inverse relationship was found between mastery and trait anxiety.

Examination of Course Effects

In the case of most variables measured in this study we were able to generate an index of course impact based upon the difference in scores from one questionnaire administration to the next. These indexes are referred to here as *change scores*. Each risk group, therefore, can be assigned a change score for every variable in relation to which it is assessed. These scores were compared across risk groups using analysis of variance (ANOVA). A statistically significant difference between risk groups provided circumstantial evidence that the course had had a differential impact upon participants according to their drinking risk status. A related approach was simply to look at differences between risk groups at baseline, using analysis of variance, then to repeat the exercise at follow-up. The expectation in many cases was that differences that existed at baseline (prior to the course) would diminish by the time of follow-

Table 5–4
Mean Change Scores in Weekly Drinking and Risk Groups

	Low	←	Risk Groups	→	High			
	1		2		3		4	
	M	F	M	F	M	F	M	F
Completers n =	17	19	34	22	9	8	5	6
Change Scores	0.35	0.29	0.22	1.24	−2.00	0.86	−9.00	−12.33

ANOVA (one-way): Males, $F = 4.465$, $P > .007$; Females, $F = 20.824$, $P > .001$.

Table 5–5
Baseline Weekly Alcohol Consumption and Direction of Change in Consumption (Sexes Combined)

Mean Weekly Alcohol Consumption[a]	Direction of Change			
	Decrease	No Change	Increase	n
0–7	15	34	22	71
	21.1%	47.9%	31.0%	64.5%
8–15	13	2	7	22
	59.1%	9.1%	31.8%	20.0%
> 15	13	0	4	17
	76.5%		23.5%	15.5%
n	41	36	33	110
	37.3%	32.7%	30.0%	100%

$\chi^2 = 29.37306$; 4 D.F.; $P > .0001$
[a]Measured in number of standard drinks.

up if the course had been successful in reaching higher risk drinkers with health messages.

Observed Changes

There were a number of apparent course-related effects. Many of them are best described in relation to the most significant change of all, which was the average weekly consumption of alcohol. This was a surprising result because little had been expected in the area of behavioral change.

If the analysis is approached first of all from the point of view of the drinking-related risk groups described previously, we can see from analysis of variance as applied to the change scores reported in tables 5–4 to 5–8 that

Table 5–6
Mean Change Scores in Tension Reduction Drinking and Risk Groups

	Low ← Risk Groups → High							
	1		2		3		4	
	M	F	M	F	M	F	M	F
n =	17	19	34	22	9	8	5	6
Change Scores	0.10	0.24	−0.10	0.29	−0.33	−0.75	−2.50	−2.33

ANOVA (one-way):   Males, F = 2.373, P > .08; Females, F = 2.546, P > .07

a number of important changes took place. (Please refer to table 5–1 for baseline values). Whether these changes are likely to have come about as a result of the course is a matter that will be taken up in the concluding section.

The actual *reduction in weekly consumption* was significant for both males and females (table 5–4). It can be seen that the effect is related to baseline consumption levels, so that the higher risk groups are those in which the greatest change took place. A somewhat different perspective on this phenomenon is provided by a cross-tabulation of baseline consumption (not controlling for body weight) and change in consumption (table 5–5). Here it can be seen that among those who at baseline consumed in excess of 15 standard drinks per week, 76.5 percent decreased their consumption while only 23.5 percent increased, with the mean change at minus 6.8 drinks per week. The same tendency, though less dramatic, can be seen among those who drank between 8 and 15 drinks per week, with the mean change at minus 1.1.

No differences were observed between those who increased or decreased their consumption in terms of age, sex, marital status, or education.

A related change took place in *tension reduction drinking* (table 5–6). While ANOVA indicates only a near significant difference between drinking-related risk groups, there is a strong correlation (0.43, P > .001) between change in consumption and change in tension reduction drinking. It appears, then, that this effect occurred across the risk groups to some extent even though it is most obvious in the high-risk group.

A similar phenomenon occurred in relation to the *definition of social drinking* where ANOVA shows a significant difference in changes between female risk groups but not between male risk groups (table 5–7). Nonetheless, the correlation between change in consumption and change in definition of social drinking is 0.60 (P > .001) for both sexes combined, that is, as people (particularly women) drink less, they tend to define moderate or social drinking in more conservative terms also. However, it cannot be assumed that the relationship between these variables is unidirectional. It is possible that the belief (definition of social drinking) affected behavior (actual drinking),

Table 5–7

Mean Change Scores in Definition of Social Drinking and Risk Groups

	Low ← Risk Groups → High							
	1		2		3		4	
	M	F	M	F	M	F	M	F
n =	17	19	34	22	9	8	5	6
Mean Change	1.21	1.80	0.30	−0.57	−3.78	−0.63	0.75	−10.00

ANOVA (one-way): Males, $F = 0.653$, $P > .58$; Females, $F = 7.505$, $P > .001$.

Note: Definition of social drinking: analysis of variance: shift in significance of differences between risk groups, pre- to post-course: summary: *Males: Pre-course*, $F = 22.266$, $P > .001$; *Post-course*, $F = 21.141$, $P > .001$. *Females: Pre-course*, $F = 22.128$, $P > .001$; *Post-course*, $F = 1.758$, $P > .17$.

Table 5–8

Mean Change Scores in Definition of Safe Number of Drinks One Hour before Driving

	Low ← Risk Groups → High							
	1		2		3		4	
	M	F	M	F	M	F	M	F
n =	17	19	34	22	9	8	5	6
Mean Change	0.13	0.13	−0.43	−0.50	−1.00	−0.62	0.0	−0.33

ANOVA: Males, $F = 1.473$, $P > .23$; Females, $F = 1.082$, $P > .37$.

Note: Definition of safe number of drinks in one hour before driving: analysis of variance: shift in significance of differences between risk groups, pre- to post-course: summary: *Males: Pre-course*, $F = 6.039$, $P > .001$; *Post-course*, $F = 3.362$, $P > .02$. *Females: Pre-course*, $F = 3.677$, $P > .02$; *Post-course*, $F = 0.558$, $P > .64$.

but only longitudinal research, employing multiple data collection points, can satisfactorily resolve this question.

The mean decrease in the number of drinks per week considered compatible with social or moderate drinking among those who defined it as more than 15 drinks per week at baseline was 8.3, a figure that is tolerably close to the 6.8 previously reported reflecting reductions in actual consumption.

The difference between risk groups that prevailed before the course tended to diminish among females by the time of follow-up (P value shifted from > .001 to > .17), while among males it remained the same (see note to table 5–7).

Table 5–9
Mean Baseline and Change Scores in Locus of Health Control and Risk Groups

	Low	←	Risk Groups	→	High			
	1		2		3		4	
	M	F	M	F	M	F	M	F
n =	17	19	34	22	9	8	5	6
Mean Baseline Scores	19.1	19.5	18.3	18.7	18.6	20.0	16.2	17.3
Mean Change Scores	−1.10	0.21	0.38	0.52	−0.67	0.50	3.00	1.67

ANOVA (one-way; change scores):   Males, F = 5.334, P > .003; Females, F = 0.546, P > .65.
Note: Analysis of variance: shift in significance of differences between risk groups, pre- to post-course: summary: *Males: Pre-course*, F = 1.531, P > .21; *Post-course*, F = 0.454, P > .72. *Females: Pre-course*, F = 2.277, P > .09; *Post-course*, F = 0.910, P > .44.

There was no significant difference between groups in relation to changes in the *definition of a safe number of drinks in one hour before driving* (table 5–8) among males or females. However, when the significance of pre- and post-course differences is compared (see note to table 5–8), we see that among females a significant baseline difference between risk groups disappeared (P > .02 to P > .64) and was reduced among males (P > .001 to P > .02).

There were differential increases in *locus of health control* according to risk group (table 5–9). Among men, a large increase in the highest risk group and a decrease in the lowest created a significant difference in change scores and a corresponding decrease in the significance of differences between all groups from pre- to post-course. Among women, there was also a "normalization" of scores from pre- to post-course evidenced by a shift from a near significant difference between groups (P > .09) to a nonsignificant difference (P > .44) (see note to table 5–9). Although the changes among women were all increases, while among men there were decreases and increases, there was nonetheless a significant inverse Pearson correlation between decrease in consumption and increase in locus of health control (−0.23, P > .03). It would appear, then, that perception of control over personal health is an important factor in bringing about or at least in reinforcing the moderation of drinking practices.

Gains in knowledge about the *signs of problem drinking* were seen among male high-risk drinkers (table 5–10). Their baseline knowledge levels on "Signs" were significantly lower than others (P > .05) but by follow-up this difference had diminished considerably (P > .64) (see note to table 5–10). Among females, the baseline picture was different in that the highest risk group was the *most* knowledgeable about the signs of problem drinking.

Table 5–10

Mean Baseline and Change Scores in Signs of Problem Drinking Recognized and Risk Groups

	Low	←	Risk Groups	→	High		
	1		2		3		4	
	M	F	M	F	M	F	M	F
n =	17	19	34	22	9	8	5	6
Mean Baseline Scores	10.3	8.6	8.0	7.1	8.2	7.4	6.2	9.5
Mean Change Scores	−0.64	1.15	1.03	2.47	1.25	0.43	2.00	−0.2

ANOVA (one-way; change scores): Males, $F = 1.230$, $P > .31$; Females, $F = 0.839$, $P > .48$.

Note: Signs of problem drinking recognized: analysis of variance: shift in significance of differences between risk groups, pre- to post-course: summary: *Males: Pre-course*, $F = 2.717$, $P > .05$; *Post-course*, $F = 0.564$, $P > .64$. *Females: Pre-course*, $F = 0.772$, $P > .52$; *Post-course*, $F = 0.121$, $P > .94$.

Although the ANOVAs do not indicate a significant change, it can be seen that females in group 2, who began with the lowest scores, increased the most in knowledge. Taken across the sample as a whole we find a correlation of −0.24 ($P > .02$) between increases in knowledge and reduction of alcohol consumption.

The message of the course in relation to the *importance of health-related behaviors or conditions* appeared to have opposite effects on men and women (table 5–11). Among men, a difference between risk groups at baseline that approached significance ($P > .08$) disappeared ($P > .42$) due largely to an increase in scores among the highest risk group. Among women, however, an initial similarity between groups diminished considerably ($P > .60$ to $P > .01$) due to an increase in scores among the *low-risk* group. This may not be of great significance, however, since the mean score on this variable was high for all risk groups both at baseline and follow-up (21.5 on both occasions in a range of 6–24).

There were no significant changes in scores associated with *factors considered very important contributors to cardiovascular disease* (table 5–12). However, an interesting "normalization" effect occurred among males so that low-risk participants' scores tended to decrease while high-risk participants' scores increased. This effect is shown to some extent in the change of F values and the accompanying probability levels ($P > .19$ to $P > .87$).

No changes of any significance were seen in relation to the *importance assigned to factors that might contribute to a personal drinking problem*. This was the case both in the sample as a whole and in the risk groups.

Table 5–11
Mean Baseline and Change Scores in Importance of Health-Related Behaviors or
Conditions and Risk Groups

	Low ←		Risk Groups		→		High	
	1		2		3		4	
	M	F	M	F	M	F	M	F
n =	17	19	34	22	9	8	5	6
Mean Baseline Scores	21.9	21.0	21.7	22.3	20.0	21.6	19.8	20.8
Mean Change Scores	−0.41	1.90	−0.76	0.10	−0.44	0.37	1.20	−1.00

ANOVA (one-way; change scores): Males, $F = 0.742$, $P > .53$; Females, $F = 1.475$, $P > .23$.
Note: Importance of health-related behaviors or conditions: analysis of variance: shift in significance of differences between risk groups, pre- to post-course: summary: *Males: Pre-course,* $F = 2.375$, $P > .08$; *Post-course,* $F = 0.958$, $P > .42$. *Females: Pre-course,* $F = 0.624$, $P > .60$; *Post-course,* $F = 4.648$, $P > .01$.

However, this obscures the fact that while some people's scores increased and others decreased—an apparently random phenomenon leading to a "canceling-out" effect—there was a pattern associated with the *increased* scores but not with those that decreased. The pattern took the form of a correlation between changes in factors and in signs of problem drinking recognized to the extent of $R = 0.30$, $P > .005$. (Further evidence of patterned change will be given in the section on factor analysis.) We should note the importance of looking for correlations among changes that in themselves appear insignificant but that in combination with others assume a relevance that exceeds that of the sum of the parts.

Similarly, no changes of any significance were seen in *mastery* according to risk groups or in the sample as a whole (mean score, baseline, 22.0). However, some of the apparently random variation in scores between first and second administrations was in fact patterned, since we find a correlation between changes in mastery and changes in the importance-of-health-behavior score ($R = 0.34$, $P > .0001$). With regard to the underlying importance of changes in mastery, however, the correlation just noted is only the tip of the iceberg as the section on factor analysis will illustrate.

Changes in the *values on personal and family health* fall into much the same category as the two preceding variables: little significance at the level of the whole sample, but patterned change in relation to certain other variables. Again, a simple correlation between changes in the value of personal health and changes in signs of problem drinking recognized ($R = 0.23$, $P > .05$) fails to capture the importance of changes in the integration of values into the fabric of everyday health beliefs and behaviors. In the framework of purely quantitative methods, this integration of values can be shown

Table 5–12
Mean Baseline and Changes Scores of Factors Considered Very Important
Contributors to Cardiovascular Disease and Risk Groups

	Low	←	*Risk Groups*	→	*High*			
	1		*2*		*3*		*4*	
	M	F	M	F	M	F	M	F
n =	17	19	34	22	9	8	5	6
Mean Baseline Scores	5.2	4.3	4.5	5.0	4.0	5.6	2.8	4.5
Mean Change Scores	− 1.2	0.4	− 0.2	0.3	0.0	− 0.4	0.8	1.2

ANOVA (one-way; change scores): Males, $F = 1.360$, $P > .26$; Females, $F = 0.604$, $P > .62$.

Note: Factors considered very important contributors to cardiovascular disease: analysis of variance: shift in significance of differences between risk groups, pre- to post-course: summary: *Males: Pre-course, $F = 1.631$, $P > .19$; Post-course, $F = 0.239$, $P > .87$. Females: Pre-course, $F = 0.646$, $P > .59$; Post-course, $F = 0.686$, $P > .56$.*

Table 5–13
Intentions to Change before Course and at Follow-Up

	Before	*After*	*Difference*
Sleep (how much, how often)	41%	43%	+2%
Work Habits (hours; schedule)	17%	43%	+26%
Smoking Habits (how often, under what circumstances)	41%	38%	−3%
Alcohol Use (how much, under what circumstances)	24%	22%	−2%
Leisure Time (how much, how it is spent)	55%	62%	+7%
Exercise (how much, how often)	70%	72%	+2%
Eating (how much, how often)	62%	66%	+4%

or suggested only through the use of multivariate statistical procedures such as factor analysis.

About 27 percent of men and 39 percent of women placed personal health in their top five values at baseline. While this picture did not change appreciably by follow-up, the relevance of the value on personal health appears to have changed (not always positively) according to factor analysis.

Table 5–13 shows the differences in rates at which respondents stated their *intentions to change* before the course and at follow-up. The most striking feature here is the stability of intentions between baseline and follow-up. Otherwise stated, participants appear to have come to the course with already formed intentions to change. The course does not appear to have influenced this picture to any great extent. Even when changes are actually imple-

Table 5–14
Relationships among Intended Changes (Follow-Up)

	Intended Change in Both	*Intended Change in Neither*
Eating and		
Exercise	67.3%[a]	13%
Leisure	56.1%	18%
Alcohol	26.6%	30%
Smoking	35.7%	25%
Work	37.0%	26%
Sleep	37.4%	24%
Exercise and		
Leisure	69%	16%
Alcohol	28%	17%
Smoking	36.7%	13%
Work	43.4%	17%
Sleep	44.9%	16%
Leisure and		
Smoking	28.9%	23%
Alcohol	28.1%	31%
Work	37.5%	25%
Sleep	41.5%	28%
Alcohol and		
Smoking	17.6%	53%
Work	20.7%	51%
Sleep	22.8%	50%
Smoking and		
Work	25.3%	41%
Sleep	23.4%	37%
Work and Sleep	33.7%	47%

[a]Percentages refer to the total respondent group at follow-up.

Table 5–15
Intention to Change and Actual Change in Alcohol Use

	Changes in Alcohol Use		
Combined Sexes	*Decreased*	*Stayed Same*	*Increased*
Intended Change (*n* = 22)	59.1%	13.6%	27.3%
No Intended Change (*n* = 64)	29.7%	39.1%	31.3%

Table 5–16
Self-Help and Care Seeking Two Months Prior to Follow-Up[a]

	Male	Female
Had a physical checkup	17 (27.9%)	25 (53.2%)
Had a fitness test	3 (5.1%)	8 (17.4%)
Sought help for an alcohol or drug problem	1 (1.7%)	1 (2.4%)
Sought help or advice for a concern about stress, depression, or other related condition	5 (8.3%)	7 (16.3%)
Joined a fitness club	6 (10.2%)	7 (16.7%)

[a]Percentages are not cumulative and sometimes overlap.

mented, the intention to change remains stated, as if to confirm the need to continue integrating the change into everyday life.

The significance of the rather large and anomalous increase in intentions to change work habits cannot be explained satisfactorily at present. It may be related to a perception that certain work habits (for example, taking the job home) or schedules (for example, shift work) are detrimental to health. Some support for this hypothesis may be drawn from an examination of the relationships among intended changes shown in table 5–14, where it is clear that people who wanted to make changes in their work habits were very likely to want to make changes in other health-related areas too, particularly exercise, eating, leisure, and sleep.

The expected relationship between intention and behavior was demonstrated with some vigor in relation to alcohol use as table 5–15 illustrates. Of those whose intention to reduce alcohol consumption was reported at baseline (pre-course), over 59 percent reported reduced consumption at follow-up, compared with 30 percent of those who made no such declaration. The course may have reinforced the resolve of those who came to the course already half-committed. Among some of those who had no stated intention of changing before the course began, the course may have instilled some resolve to change. However, it is more likely that random variations in drinking over time account for most of the decreases and increases among "non-intenders."

Respondents were asked whether they had engaged in any *new health-related activities or behaviors* in the two months preceding administration of the follow-up questionnaire. The results are shown in table 5–16. In addition, 25 percent of respondents noted in an open-ended section of the questionnaire that they were trying to make changes in their eating habits and 31 percent said they were trying to get more exercise. However, these figures exceeded baseline figures by only 5 percent and 13 percent respectively. It is impossible to tell from these data alone if the course provided these participants with any clearer direction about the kinds of changes they could effectively make.

Care seeking and self-help were related in some instances to drinking risk levels and within this context, to sex. Thus, 83 percent (or five out of six) high-risk women reported that they had had physical checkups in the last two months in comparison with the 53 percent average rate among women in this sample. Among men, the trend was reversed so that none of the high-risk drinkers had taken physical checkups. This may have been a function of the fact reported earlier that, at baseline, high-risk males tended to deny (or simply did not have) health problems. Among males, nondrinkers had the highest reported rates of physical checkups (41 percent), followed by low-risk drinkers (27 percent), and medium-risk drinkers (12 percent). Women generally sought physical checkups at almost double the rate of men.

Fitness tests—rare events in any case—were not taken by high-risk men or women. However, two of the seven medium-risk women took such tests even though no males in this category did so. Assistance for alcohol problems was sought by one high-risk woman and one medium-risk man.

Assistance or advice for stress, depression, and related problems was sought by men and women in all drinking risk categories, although a greater tendency to do this was found among high-risk drinkers. It is interesting that among males who sought help for stress, depression, or a related condition 80 percent also cut down on their alcohol consumption, while among females 57 percent cut down. However, while none of the males increased their consumption, 43 percent of the females seeking help for stress, depression and other conditions did so. The seeking of help was also related to having physical checkups. Eighty-three percent of those who sought help also had checkups, compared with 30 percent among those not seeking help.

Joining a health club was more likely among medium-risk men (29 percent or two out of seven) than among other groups where the average was just below 10 percent. Among women, no high-risk drinkers joined health clubs and of the remainder it was somewhat more likely that nondrinkers and low-risk drinkers would do so than medium-risk drinkers. Males joining health clubs were more likely to reduce their alcohol consumption (50 percent decreased, while 17 percent increased) but females were more likely to increase (14 percent decreased, while 71 percent increased). These differences are clearly related in part to baseline consumption levels that were lower in the case of women who joined health clubs.

The role of *social support* was explored in relation to its impact upon those who made changes in their alcohol consumption. The general picture is shown in table 5–17. It appears that whether support is seen as coming from intimate partners or from any source at all (friends, workmates, rela-

Table 5–17
Social Support and Changes in Alcohol Use

	Changes in Alcohol Use		
Source of Support	*Decreased*	*Stayed Same*	*Increased*
Spouse/Partner Support *n* = 20 (19%)	55%	5%	40%
No Such Support *n* = 84 (81%)	32%	39%	28%
Support from Anyone *n* = 33 (32%)	61%	6%	33%
No Such Support *n* = 70 (68%)	26%	44%	30%

tives, and so on, but also including intimates), it is influential. With regard to spouse/partner support, however, the likelihood of *decreases* in alcohol use is somewhat less than in the case of support from all sources combined. Indeed, it is interesting to note that spouse/partner support is also quite likely to be associated with *increases* in consumption. These increases, when they occur, tend to be larger than in cases where no support for change is recorded. Thus, among those who had support but increased their consumption, 63 percent did so by more than three drinks a week compared with 38 percent among the unsupported. (It is possible, of course, that support or encouragement could have been given for *increased* consumption if baseline consumption was *low*. While these observations do not provide the basis for definitive conclusions, we draw attention to the possibility that social "support" may be double-edged.) The role of support in the modification of drinking or any other health-related behavior is worthy of further investigation given the abandonment of assumptions as to its positive nature.

The Structure of Health Attitudes and Beliefs Surrounding the Use of Alcohol: Factor Analysis

In presenting the data reported so far we have suggested that certain general health practices, beliefs, attitudes, and values co-vary to some extent with more specific practices, beliefs, and attitudes related to the consumption of alcohol. We have shown too that changes in the one set of beliefs, attitudes, and behavior (the general health constellation) are sometimes significantly correlated with the other set (the specific alcohol constellation). In both cases we implied that such changes were a result of the healthy life-styles course, "Take Charge."

With regard to the latter claim it must be pointed out that the evidence to date is circumstantial. Further tests of the intervention, utilizing quasi-experimental designs must be implemented in order to render a more definitive body of evidence. Nevertheless, there are promising leads in the present body of data concerning what we might expect with regard to relationships between the general health constellation and the specific alcohol constellation and how both sets respond to the kind of health messages contained in the Take Charge course. So far these relationships have been explored through analysis of variance. As applied here, this approach is essentially bivariate. In this section we take the exploration a step further, employing factor analysis to determine aspects of the structure underlying the constellations referred to above. We present the results in terms of factor structures for both pre-course and post-course data sets. The participant samples are identical in both cases. Consequently, we should be able to think about differences in factor structure from before to after the course as to some extent indicators of changes brought about by the course. Cause–effect relationships, where postulated, are presented in a suggestive rather than a definitive manner.

Data are presented for males and females separately. The form of analysis is identical in all cases: the initial factors are rotated to an oblique solution. The factor structures rather than the patterns are reported here. Consequently, one may read the loadings as correlations with the postulated underlying factors. The titles ascribed to the factors are attempts to express the investigators' main interpretation of the cluster.

Loadings below 0.20 (approximately 4 percent of the variance of a variable) are reported only where the distribution of variance across factors is of significance—that is, the variable appears on more than one factor. Similarly, factors with eigenvalues of less than 1.00 are reported because of their relationship to other factors in which the same or similar variables appear. In oblique rotations the resulting factors are allowed to be correlated. Thus, such correlations are reported where they appear.

The *order* of the factors is not necessarily nor even likely to be a reflection of the prevalence of the behaviors, attitudes, and beliefs found in the sample. Rather, order reflects the interrelatedness of such behaviors, attitudes, and beliefs. The greatest interrelatedness could occur among adherents to a minority view, so that the structure of this group's beliefs and practices would dominate the pattern of relationships in the data so as to generate them as elements in the first factor, that is, the factor that explains the greatest single amount of variance among the variables.

Factor Analysis of Variables before Course: Males.

> Factor 1: *Correlates of Committed Heavy Drinking with No Intended Changes*
> (percent variance explained, 37.9%; eigenvalue 3.142)

	Loadings
Mastery	0.24
Importance of health behaviors	− 0.20
Total alcohol per week	0.88
Drinks per typical occasion	0.27
Signs of problem drinking recognized	− 0.33
Tension reduction drinking	0.20
Excessive drinking (personal) contributors	− 0.20
Definition of social drinking	0.96
Definition of safe drink/drive criterion	0.26
Importance of factors contributing to heart disease	− 0.20

Comment on Male Factor 1. The dominant feature of this factor is the relationship between actual weekly consumption of alcohol and the definition of what constitutes acceptable social drinking. There is little recognition of signs that indicate the emergence of a drinking problem in others or of factors that might contribute to a personal drinking problem. The pattern of behavior, attitudes, and beliefs described by this factor includes tension reduction drinking but not to any significant degree. Similarly, a variety of behaviors and conditions associated with well-being and the prevention of illness (including cardiovascular disease) are not given much importance. The amount of alcohol that can be safely consumed in one hour prior to driving is liberally defined in this factor, although the amount of variance explained is less significant here than in other factors. Because of the relative absence (compared with other factors) of variance in "drinks per typical occasion" and the emphasis on large overall amounts, we suggest that this factor is best typified as one describing chronic heavy drinking rather than acute episodic drinking. The greatest amount of variance in "drinks per typical drinking occasion" is involved in factor 5, which in fact correlates with factor 1 to the extent of 0.27.

Note the slight involvement of mastery in factor 1. Although the amount of variance involved is small, it is interesting that higher mastery scores should be represented at all. The intuitive feeling about the kind of men described by this factor is of a sanguine group of chronic heavy drinkers who do not feel that their health or well-being is threatened by their alcohol consumption and who downgrade or deny the importance of healthy behavior in terms either of present well-being or future illness.

Factor 2: *Intimation of Change among Episodic Tension Relief Drinkers*
(percent variance explained, 19.3%; eigenvalue 1.594)

	Loadings
Value placed on family health	-0.23
Mastery	-0.32
Locus of health control	0.33
Sum of attempted changes	0.37
Drinks per typical drinking occasion	0.41
Signs of problem drinking recognized	0.20
Tension reduction drinking	0.76
Excessive drinking (personal) contributors	0.31

Comment on Male Factor 2. This factor appears to describe a situation in which certain men who drink to relieve tension or to otherwise deliberately modify how they feel are aware that their drinking may be causing or about to cause them problems. The pattern of drinking is probably episodic because the factor includes amounts consumed per occasion but not total amounts per week. In addition, these men have tried to make certain changes in their life styles during the past year. Note that these changes are not specific to alcohol use, though they include it. The score sum of attempted changes, it will be recalled, is derived from the addition of recorded attempts at change in eating, sleeping, exercise, work, recreation, smoking, and alcohol-use practices. Men associated with this factor seem to feel that they have a good chance of success if they decide to make health-related changes (locus of health control) but this sense of self-efficacy is not a general one since mastery scores are low. This negative correlation between locus of health control and mastery is manifested in the total sample as -0.28 at baseline and as -0.35 at follow-up. Thus, if anything, the inverse relationship became stronger. This was probably more a function of increases in scores on locus of health control than decreases in mastery, although both occurred. Such a negative correlation is not unexpected in light of the predictions of various investigators such as Beck, McAlister, Bandura, and Fishbein discussed earlier in this chapter. Their common point is that self-efficacy is a concept best operationalized in relation to specific behaviors and that general self-efficacy is likely to be a poor predictor in itself of perceived competence in specific areas. Consequently, it should not surprise us to see that a robust sense of mastery can coexist with a flaccid sense of efficacy with regard to health matters, nor indeed that such a disparity might even be increased as a result of exposure to certain types of health messages.

Factor 3: Moderation and Concern among the "Worried Well"
(percent variance explained, 16.0%; eigenvalue 1.321)

	Loadings
Mastery	0.22
Importance of health behaviors	0.57
Locus of health control	−0.88
Sum of intended changes	0.20
Sum of attempted changes	0.32
Total alcohol per week	−0.31
Signs of problem drinking recognized	0.20
Tension reduction drinking	−0.22
Importance of factors contributing to heart disease	0.55
Definition of social drinking	−0.28
Definition of safe drink/drive criterion	−0.22

Comment on Male Factor 3. This factor can be seen as describing a relatively high level of health awareness (importance of health behaviors and factors contributing to heart disease) coupled with moderate alcohol-related attitudes and practices. In addition there is evidence of attempts to improve personal health (sum of attempted changes) and of intentions to continue this endeavor (sum of intended changes). In spite of this, scores on locus of health control and mastery load in opposite directions. Consistent with observations made in relation to male factor 2, it appears that among those who may be described as "health moderates," as in factor 3, the impetus to improve or reinforce healthy practices is associated with a notably below-average sense of personal control over health but with a somewhat above-average sense of general self-efficacy. However, among those whose drinking behavior, at least, is immoderate (as in factor 2) we find that the impetus to change (assuming it is genuine) is associated with a somewhat above-average sense of personal control over health but with a somewhat below-average sense of general self-efficacy. Whether these differences are psychologically meaningful is a question that cannot be answered satisfactorily by reference to the data at hand. However, we might speculate that in the case of people (men, at least) who feel they have less control over their lives (lower mastery scores), the identification of an area in which they *could* make changes—drinking, for example—might be reflected in a sense of empowerment (higher scores on locus of health control) in the health area but not in more general areas. Conversely, people who generally feel control over their lives, and who are basically healthy may see health as a particularly vulnerable state (lower scores on locus of health control) and one that requires considerable effort to maintain (higher scores on importance of health behavior). We refer to such people as the "Worried Well." Thus, in so far as locus of health control refers to the perceived difficulty of making changes and to the controllability of risk factors, it may be that experience leads to more cautious attitudes on the part of the "Worried Well" who contemplate a wide spectrum of behaviors as constituting "well-being." Less moderate people, on the other hand, may be more sanguine about the chances of controlling their own health, partly because they may think about it in less global terms (perhaps focusing on drinking as the problem) and partly, perhaps, because they are more optimistic people generally. Clearly, a definitive test of these hypotheses is outside the realm of this investigation.

Factor 4:   *Concern among Episodic Heavy Drinkers*
(percent variance explained, 10.3%; eigenvalue
0.848)

	Loadings
Value on personal health	−0.20
Value on family health	−0.20
Mastery	0.20
Sum of intended changes	0.62
Sum of attempted changes	0.33
Total alcohol per week	−0.20
Drinks per typical drinking occasion	0.34
Signs of problem drinking recognized	0.39
Importance of factors contributing to heart disease	0.24
Definition of safe drink/drive criterion	0.46

Comment on Male Factor 4. This factor appears to describe men who drink large amounts on occasion but not a huge amount overall and who cherish the belief that one may drink a lot and still drive safely. At the same time there is an awareness of the signs of problem drinking and to some extent an understanding of heart disease risk factors. They have made attempts during the past year to make health-related changes, and they express the intention to continue this process across the spectrum of health-related behaviors. This pattern of knowledge, attitudes, and behavior includes a view of self as being in control of events, though not to any exceptional degree, but it does not include tension reduction drinking. It is difficult to deduce what the impetus for change is among the men described by this factor since personal and family health are little valued. However, the paradox of concern about heart disease—if we may deduce "concern" from "ascribed importance of risk factors"—may supply a clue to the desire to make health related changes.

Factor 5:   *Episodic Tension Relief Drinking*
            (percent variance explained, 7.3%; eigenvalue
            0.605)

	Loadings
Total alcohol per week	0.39
Drinks per typical drinking occasion	0.83
Signs of problem drinking recognized	− 0.52
Tension reduction drinking	0.41
Importance of factors contributing to heart disease	− 0.46
Definition of social drinking	0.31
Definition of safe drink/drive criterion	0.24
Excessive drinking (personal) contributors	− 0.26

Comment on Male Factor 5. The high loading on drinks per occasion defines this factor, although we see that overall amounts consumed are relevant also. The episodic pattern of consumption is associated with a pursuit of relief from tension and is to some extent reinforced by beliefs about acceptable standards for social drinking and drinking and driving. There is a low awareness of the signs of problem drinking and of the risk factors for heart disease. There is neither any evidence of attempts to bring about health-related changes nor of any intention to do so. As noted already, factors 5 and 1 are correlated to the extent of 0.27. Both describe patterns of behavior and thinking in which the prospect of change is conspicuous by its absence.

Factor Analysis of Variables after Course: Males. In reviewing the factor analysis of the follow-up data it is important to bear in mind that the participants whose responses are shown here had undergone exposure to a health-oriented course. We have shown already that certain changes took place between pre- and post-course questionnaire administrations. We might anticipate, therefore, that a certain amount of restructuring of behavior, attitudes, and beliefs would reveal itself if we compare relationships within the data at follow-up with relationships observed at baseline. This restructuring is apparent in the context of the factors emerging from the follow-up data.

Factor 1 (Post):   *Attempted Changes among Episodic*
                    Tension Relief Drinkers
                    (percent variance explained, 37.7%;
                    eigenvalue 3.137)

	Loadings
Value on personal health	−0.28
Locus of health control	0.29
Sum of attempted changes	0.51
Total alcohol per week	0.20
Drinks per typical drinking occasion	0.70
Tension reduction drinking	0.59
Definition of social drinking	0.27
Definition of safe drink/drive criterion	0.63

Comment on Male Factor 1 (Post). This factor appears to describe the correlates of episodic tension relief drinking as structured at the time of follow-up. The elements of the factor bear some resemblance to baseline factors 2, 4, and 5. The essence of this constellation is one of continued bouts of heavy drinking for purposes of seeking relief from tension or changing mood. Personal health is relatively undervalued, while at the same time there is a tendency to believe that the individual has control over it. A strong belief persists that one may drink large amounts and still drive safely. In spite of this, there have reportedly been attempts at making health-related changes. Clearly, these attempts had little or no relevance to drinking or its associated supporting beliefs and attitudes. In fact, one might speculate that this factor is describing *negative* reinforcement of behavior—that is, a situation in which attempts were made at change that failed, leading to a reinforcement of excessive drinking practices. Note that in distinction from factor 2 in the first administration, factor 1 at follow-up does not involve variance in "signs of problem drinking recognized" or in "excessive drinking (personal) contributors." This suggests a diminution or even denial of the importance of warning signs in relation to problem drinking. Factor 1 (post) should be compared with factor 4 (post) with which it has a negative correlation (−0.15). As reported below, factor 4 (post) seems to be a mirror image in some respects of factor 1 (post). It appears to describe the results of *positive* reintegration of behavior and beliefs among episodic drinkers who have *reduced* their consumption. However, we may be seeing evidence here of behavioral and attitudinal polarization, that is, a situation in which the health messages in the course cause high-risk participants to consolidate their behavior and supporting beliefs in both positive and negative directions as defined in relation to the course objectives. At the same time, we must recall that the overall picture is one of a significant mean reduction in alcohol consumption. We have seen too that some people *increased* their consumption, albeit a minority of the high-risk drinkers. What may be of concern is that the subgroup of higher risk drinkers who did not change their consumption at all may have been reinforced in their belief that they *could not* change in spite of a protestation that health changes were within their control.

At the very least, the consolidation of variance between the two questionnaire administrations in relation to episodic tension relief drinking is a disturbing trend. We should note, however, that in addition to the course a strike intervened between the two test administrations. In terms of impact upon drinking the Take Charge course and the strike were probably competing influences, since it is believed that consumption tends to increase among workers while unemployed (Smart 1979), although Giesbrecht, Markle, and Macdonald (1982) found the opposite to be true in the case of a long strike at INCO (a large mining company in northern Ontario). However, *length* of unemployment probably has a bearing on consumption. The strike in our case was of relatively brief duration (five weeks) in comparison with the INCO strike (eight and a half months). In long strikes, one might hypothesize an initial increase in consumption associated with extra leisure time and tension reduction followed by a later decrease associated with financial hardship. In the case of the manufacturing plant that is the subject of this report, no one really expected a long strike so the financial hardship factor may not have been crucial. In addition, it is essential to note that the follow-up was conducted more than one month after the strike had ended. Questions about consumption were directed at the month prior to questionnaire administration.

	Loadings
Factor 2 (Post): Reinforcement of Moderation among the Worried Well (percent variance explained, 19.1%; eigenvalue 1.593)	
Value on personal health	0.22
Mastery	0.54
Importance of health behaviors	0.84
Locus of health control	−0.53
Sum of intended changes	0.50
Sum of attempted changes	0.25
Drinks per typical drinking occasion	−0.21
Signs of problem drinking recognized	0.33

| Importance of factors contributing to heart disease | 0.41 |
| Definition of safe drink/drive criterion | −0.22 |

Comment on Male Factor 2 (Post). This factor resembles in most respects factor 3 at baseline. Value on personal health appears at follow-up while more variance in mastery and importance of health behaviors is involved. The negative relationship between mastery and locus of health control has become stronger. The dynamics underlying this relationship were explored under the discussion of baseline factor 3. The amount of variance explained by this factor at follow-up is somewhat greater than at baseline, suggesting if anything a reinforcement of the behavior, beliefs, and attitudes described by the latter. This can be seen as a result that is consistent with the objectives of the course.

Factor 3 (Post): Contented, "Committed" Heavy Drinkers
(percent variance explained, 17.0%;
eigenvalue 1.412)

	Loadings
Intended changes	−0.51
Attempted changes	−0.24
Total alcohol per week	0.86
Drinks per typical drinking occasion	0.44
Signs of problem drinking recognized	−0.31
Definition of social drinking	0.93
Definition of safe drink/drive criterion	0.20
Tension reduction drinking	0.20

Comment on Male Factor 3 (Post). This factor echoes factor 1 at baseline. The main differences are in the amount of variance explained (17.0 percent at follow-up compared with 37.9 percent at baseline) and in the probably related disappearance of correlates in the form of factors contributing to a personal drinking problem and risk factors related to heart disease. Therefore, while heavy drinking is still supported by undesirable definitions of what constitutes social drinking and safe levels of drinking before driving, it is no longer supported by ignorance about factors that might contribute to a personal drinking problem or to cardiovascular disease. Unfortunately, it appears that a dissonance remains between this assumed new knowledge or awareness and its application to definitions of social drinking and drinking and driving safety limits. Even so, the acquisition of some new information about life style factors that may contribute to physical problems, even if the personal relevance of this information has not yet been internalized, should be considered a step toward achievement of the course's objectives.

Factor 4 (Post): (Increased) Health Awareness among
                  (Former) Episodic Drinkers
                  (percent variance explained, 12.6%;
                  eigenvalue 1.049)

	Loadings
Value on personal health	0.37
Sum of attempted changes	−0.38
Total alcohol per week	−0.24
Drinks per typical drinking occasion	−0.34
Signs of problem drinking recognized	0.56
Tension reduction drinking	0.22
Importance of factors contributing to heart disease	0.31
Excessive drinking (personal) factors	0.67

Comment on Male Factor 4 (Post). As noted under the discussion of factor 1 (post) above, factor 4 (post) is negatively correlated with the former to the extent of −0.15. This apparently insignificant relationship is a clue to the meaning of factor 4, however, in that it suggests evidence for a positive effect of the course that mirrors the relatively negative effect sketched by factor 1. In ascribing to factor 4 the title "Increased Awareness among Former Episodic Drinkers" we have placed the terms "increased" and "former" in parentheses to reflect the speculative nature of the interpretation. The argument in favor of this meaning is that we have evidence of the kind of changes suggested by factor 4 (post) from the bivariate and risk group analysis data presented earlier. We might expect to see structural evidence of these changes in the form of a factor or factors showing reintegration of beliefs, attitudes, and behaviors. Factor 4 (post) then, can be seen as reflecting the results of certain correlated changes consistent with the objectives of the course. According to this interpretation, the low levels of alcohol consumption involved in factor 4 (post) reflect *decreases* that occurred between baseline and follow-up, while the high scores on recognition of problem drinking signs, excessive drinking (personal) factors, and to some extent heart disease factors represent evidence of *increases* in knowledge and awareness related to these areas. Similarly, the appearance of the value on personal health as positively related to factor 4 (post), compared with the negative relationship in factor 4 baseline, suggests both the *integration* of this value into health-related beliefs and behavior as well as repositioning of this value relative to others.

The negative relationship of sum of attempted changes to this factor suggests that change in alcohol consumption among some (episodic) drinkers is an event (or series of events) that takes place as a result of focusing upon this behavior, while attempts to make other health-related changes are put aside or not considered. This focusing, however, should not be taken as evidence that changes in alcohol consumption can be accomplished without modification of supporting beliefs and attitudes. Factor 4 (post) does, in fact, depict a variety of these supporting beliefs and attitudes, for example, signs of problem drinking recognized, importance of factors contributing to heart disease, factors believed to contribute to a potential personal drinking problem. As noted, the (new) behavior is supported also by a reinforced value placed upon personal health.

Factor Analysis of Variables before Course: Females.

Factor 1: *Correlates of Heavy Drinking (with Intended Changes)*
(percent variance explained, 42.2%; eigenvalue 4.187)

	Loadings
Value on personal health	− 0.16
Value on family health	− 0.14
Importance of health behaviors	0.26
Locus of health control	0.41
Sum of intended changes	0.48
Total alcohol per week	0.89
Drinks per typical drinking occasion	0.88
Tension reduction drinking	0.86
Excessive drinking (personal) factors	0.46
Definition of social drinking	0.89
Definition of safe drink/drive criterion	0.55

Comment on Female Factor 1. There are several differences between factor 1 for females and factor 1 for males at baseline. While it is clear that the same support for heavy drinking exists for women as for men in the form of how social drinking and drinking and driving standards are defined, women's heavy drinking differs in that it is more obviously associated with tension reduction, and there is an awareness of factors in one's personal life that could lead to excessive drinking. Similarly, there are signs that life-style changes are intended, and a tendency to believe that a healthy life style is important, although neither personal nor family health are highly ranked as values. In addition, there is a belief that changes can be made and that they will make a real difference to health. Among men whose behavior, beliefs, and attitudes are described by male factor 1, these health correlates bear a negative relationship to the dimension intimated by the factor. (It is important to remember in relation to drinking by women in this sample that the mean rate of consumption is lower than among men.)

Factor 2:   Correlates of the Value on Family Health
            (percent variance explained, 22.6%; eigenvalue
            2.244)

	Loadings
Value on personal health	0.19
Value on Family health	0.83
Mastery	0.14
Importance of health behaviors	0.15
Locus of health control	0.14
Sum of intended changes	0.56
Sum of attempted changes	0.70
Drinks per typical drinking occasion	0.26
Signs of problem drinking recognized	0.42
Excessive drinking (personal) contributors	0.16

Comment on Female Factor 2. This factor appears to describe the correlates of placing a high value upon family health among women. The principal correlates appear to be a history of attempting to make health-related changes and a readiness to make others. This constellation of values, attitudes, beliefs, and behavior involves some seemingly moderate episodic drinking (again, defined relative to the sample norm) but also a significant awareness of the signs of problem drinking. Other health-related variables with low loadings on the factor are shown here to indicate that their involvement, though weak, is in the direction of placing a positive premium on the importance of health behaviors and upon the value of personal health.

Factor Analysis of Variables after Course: Females.

Factor 1 (Post):   Correlates of Heavy Drinking (without
                   Intended Changes)
                   (percent variance explained, 35.5%;
                   eigenvalue 3.336)

	Loadings
Value on personal health	0.56
Value on family health	−0.84

Total alcohol per week	0.58
Drinks per typical drinking occasion	0.22
Signs of problem drinking recognized	0.20
Tension reduction drinking	0.63
Excessive drinking (personal) contributors	0.29
Definition of social drinking	0.68
Definition of safe drink/drive criterion	0.70

Comment on Female Factor 1 (Post). In some respects this factor resembles factor 1 at baseline. However, several changes are evident. The value on personal health is prominent at follow-up but not at baseline. However, there is no longer any involvement of sum of intended changes, suggesting either that the changes intended at baseline have been implemented or that the intended changes have been abandoned. An argument in favor of the former interpretation is that the amount of variance in the drinking related variables is less here than in baseline factor 1 suggesting that follow-up factor 1 is a record of associated changes that took place between the two administrations. If this is the case it would appear that positive reassessment of the value on personal health and negative reassessment of the value on family health were involved in the decline in consumption. In the process, however, it seems that an erosion took place with regard to feelings of self-efficacy in relation to health (locus of health control), since this variable does not appear in follow-up factor 1. Definitions of social drinking and safe drinking and driving standards were not significantly affected. We cannot suggest any good reason at present for the negative premium placed on the value of family health. Apparently, it declines in significance in some sort of lock step with an increase in the premium placed on personal health. One might speculate that this phenomenon is an indicator of the need for women to "make room" for their own health concerns by relatively devaluing family health concerns. We do not know from the data at hand whether the "families" within the respondents' phenomenological realities are spouses, live-in partners, aging relatives, children, or other dependents.

Factor 2 (Post):   *Reinforcement of Health Awareness and Health Anxiety*
(percent variation explained, 22.2%; eigenvalue 2.091)

	Loadings
Value on personal health	0.26
Importance of health behaviors	0.56
Locus of health control	−0.92
Total alcohol per week	−0.21

Drinks per typical drinking occasion	−0.57
Signs of problem drinking recognized	−0.24
Tension reduction drinking	−0.18
Importance of factors contributing to heart disease	0.48
Definition of safe drink/drive criterion	−0.36

Comment on Female Factor 2. Baseline factor 2, which dealt with the correlates of the value of family health including intended and attempted changes, has given way to this factor in which the family health value is not to be found but where a low to moderate involvement of the value on personal health has become more prominent. It is unclear whether the apparent competition between personal and family health values is a reflection of the process speculated upon in the discussion of factor 1 (post). Most noticeable in this factor is the negative correlation of locus of health control and the positive correlation of importance of health behaviors. This suggests that an increase in the importance ascribed to healthy living (both the process and the outcome) is gained at the expense of a diminution in the individual's sense of self-efficacy with regard to healthy living. In the case of women this trend appears to have been exacerbated as a result of the course since the simple correlation between locus of health control and importance of health behaviors went from −0.07 at baseline to −0.41 at follow-up. With men, a baseline correlation of −0.45 remained constant at follow-up. This pairing of increased ascription of importance to healthy living and decreased feeling of being able to do anything about it suggests a state of "healthy anxiety" similar to the condition described elsewhere as characterizing the "worried well." This apparent restructuring of attitudes and beliefs is accompanied by low levels of alcohol consumption, a conservative definition of safe drinking and driving standards and a strong tendency to be able to identify factors that contribute to heart disease. Since these variables did not appear in factor 2 at baseline, we might conclude that their appearance at follow-up is evidence of changes that took place. Since we know from evidence already presented that reductions in consumption of alcohol and gains in knowledge took place we should not be surprised to see these changes interrelated structurally.

The Impact of "Take Charge": Summary

The overall purpose of Take Charge was to reinforce life-style supports for moderate, low-risk consumption of alcohol and to reduce high-risk consumption. This purpose was to be achieved through influencing the health-related beliefs and attitudes thought to support immoderate alcohol use and through promoting an awareness of drinking in the context of lifestyle as a system of interdependent behaviors, beliefs, and attitudes. The primary impact was expected to occur upon intentions to change behavior rather than upon actual behavior. The surprise results were that participants came to the course with many intentions to change already formed and that a noteworthy proportion of them actually made changes, particularly in the area of alcohol use and exercise. The reduction in alcohol use was supported to a significant degree by changes in knowledge and beliefs related specifically to drinking (the "alcohol constellation") but also by a cluster of more general health-related attitudes and beliefs (the "health constellation"). There is evidence that an integration took place within and between these two constellations of knowledge, beliefs, attitudes, and behavior. In addition, it appears that the value upon health became more salient for some people. This process may perhaps be referred to as a breaking down of dissonance between what is valued and what is actually done.

It is important to note, however, that not all the reintegration observed in this study can be considered in a positive light. There was some evidence from the factor analysis that "undesirable" behaviors, beliefs, and attitudes were also reinforced in some cases, for example, the impact of the course upon episodic tension-relief drinkers appears to have been one of polarization. Some such drinkers improved while others deteriorated. However, the overall statistical picture suggests that the former situation was more common than the latter. Among certain women, too, there appears to have been a negative effect in that sometimes self-efficacy with regard to health was eroded rather than reinforced to the point where attempts at change may have been abandoned.

The experience of this evaluation suggests that audience information should be specifically targeted with reference to health needs. Age, sex, and education all seem to influence not only basic belief and attitude structures but also related health needs. We do not propose that audiences should be physically split up, but rather that health messages be constructed in such ways that every segment's needs are addressed to some extent.

Clearly, even though people came to the course with stated intentions to change, many found it difficult to act upon these intentions. Some self-help behavior was observed in the form of joining health clubs, going for fitness tests, getting medical examinations, and even seeking help for emotional problems. However, many were not able to take the initial steps toward change recommended in the course.

Much was learned here with regard to the importance of self-efficacy in making and not making health-related changes. Future course designers might bear in mind the negative relationship between general feelings of control over life (mastery) and feelings of self-efficacy with regard to specific areas such as health practices. On the other hand, *minor* reductions in sense of self-efficacy with regard to specific health behaviors need not be untoward experiences for the individual since, as noted earlier, too much confidence can be associated with inappropriate risk taking.

It was not always clear, however, what led to reductions in self-efficacy as related to health (locus of health control) among some of the participants in the course. If the reduction implied a greater realism about what could be affected and what not, we might consider such a phenomenon benign. If it implied, however, that the prospect of any health change was too overwhelming, we need to consider ways of increasing the perception that small changes are disproportionately important. It is conceivable that presenting eating, smoking, exercising, drinking, sleeping, and so forth in the context of an interdependent *system* of health-related behaviors made some people feel that they would not be able to change any part of it at all, rather than feeling that any small part that they did change would ultimately influence every other area. An implication is that more attention should be paid to demonstrating the positive aspects of the interdependence of health behaviors so as to reinforce rather than sabotage expectations of efficacy in making changes.

It is conceivable that by introducing novel components into HPPs of the type discussed here, the issue of empowerment or self-efficacy could be more fully addressed. For example, the use of self-help groups focusing perhaps on specific topics of common concern might aid in the effort to provide practical ideas for change and simultaneously to generate support for such changes. These groups might "spin off" from HPPs of various kinds, provided that the leaders or co-ordinators are trained in the facilitation of groups and are sensitive to the diverse needs of HPP participants. This should not be read to imply that professionals become involved with self-help groups on a direct service basis—that would be a contradiction in terms. However, they might serve as resource consultants after the initial facilitation phase. Self-help groups and support groups are not, of course, without their problems. Some of the issues involved in their establishment and maintenance are touched upon in the last chapter of this book.

Given the prominence assigned to linkages between HPPs and EAPs in the preceding chapters and elsewhere in this book, readers may have been struck by the EAP-like sequelae of the Take Charge program in the form of people seeking help for stress, depression, anxiety, alcohol dependence, and emotional problems. In fact, 8.2 percent of the study population sought help in one or more of the areas just listed. While we cannot be confident that all of this care seeking was related to this course, it seems likely that much of it was catalyzed by the program. Further investigation is required to determine what the "normal" rate of care seeking would be without such a stimulus. Meanwhile, we propose the hypothesis that programs like Take Charge are capable of generating help-seeking behavior that, in the presence of a well-developed EAP, could result in the effective treatment of problems at presumably an early stage. This may turn out to be a major element of the symbiosis between HPPs and EAPs.

6

The Impact of Stress Management Programs on the Mental Health of Workers: Two Case Studies

with Eli Bay and James Simon

I n this chapter we present descriptions and evaluations of two quite different stress management programs, both of which were designed to meet the needs of employed people. Consistent with the theme of this book, these case studies are presented on the assumption that the programs to which they refer are relevant to a comprehensive approach to mental and physical health in the workplace. The results suggest that stress management can be an extremely important ingredient in such a comprehensive approach.

These programs, however, when carried out in the workplace, have received scant attention from the scientific community in spite of their popularity (Schwartz 1982). Nonetheless, three studies are conspicuous in the attentiveness of their efforts to evaluate various stress management programs among employed populations, namely those of Peters, Benson, and Porter (1977a; 1977b), of Carrington and associates (1980), and of Freidman, Lehrer, and Stevens (1983). Numerous other studies of stress management or of the more specific relaxation training do exist (see Hillenberg and Collins [1982] for a review) but the delivery of such interventions in or through the workplace adds a dimension to such training that warrants attention in its own right. With regard to the impact of stress management or relaxation training upon the consumption of alcohol and drugs among normal employed populations, nothing has been published in the scientific literature at all.

Case Study 1. Evaluation of "Beyond Stress": A Relaxation Training Course

In the first half of this chapter we report on the manner in which a relaxation-training course offered by a commercial firm affected stress, strain, and the use of alcohol, drugs, and coffee among self-selected participants from workplace settings.

The course itself was made available to companies and individuals by the

Toronto firm "Relaxation Response Ltd.," a title borrowed from the earlier book of the same name (Benson 1975). The principal of this firm, who was also the course instructor, Eli Bay, requested the first author to undertake an evaluation of the course, since he needed objective information on the basis of which to refine or modify the content of the sessions. This relationship was governed by a contract between Relaxation Response Ltd. and the Addiction Research Foundation of Ontario.

The first author, Shain, (hereafter referred to as the investigator) undertook the evaluation, since courses of this type are becoming quite common in the context of workplace HPPs (Danielson and Danielson 1980; Fielding and Breslow 1983), and yet their impact is not widely understood, as implied earlier. The research mandate of the investigator involves the study of any and all interventions carried out in the workplace that have a reasonable probability of affecting the mental health of workers and the incidence and prevalence of problems associated with alcohol and drug use. Since a connection between stress and such problems has been long postulated (for example, Cappell 1975), the present evaluation appeared warranted as a test of the value of certain stress reduction methods offered under commercial auspices in the prevention of excessive alcohol and drug use. It should be emphasized that the value of the course is not established or disproven simply in relation to its accomplishment of objectives in the domain of substance abuse. Clearly, effective stress reduction methods can be expected to have positive, general effects upon the present and future health, well-being and productivity of those who practice them. These effects are considered here within the bounds of a quasi-experimental evaluation carried out over a period of approximately six months. A more general overview of the role of stress in future morbidity and mortality is given in chapter 2.

The Course

It is important to note that the course (entitled "Beyond Stress") was not promoted as a total stress management package. It was offered as a smorgasbord of relaxation techniques that represent part, albeit an essential part, of a comprehensive approach to stress management. Further, it was not delivered as a clinical intervention in that the sessions were carried out in group settings with no diagnosis of individual conditions and no attention paid to matching of participant needs with available techniques for relaxation except in so far as such a process occurred through informal personal dialogue with the course instructor. Consequently, the mode of intervention was education- and training-oriented rather than treatment-oriented. It is possibly this orientation that accounts for the popularity of such courses; few or no stigmas appear to be associated with involvement in a training program, whereas "treatment" tends to imply the existence of psychological or medical problems. In addition, of course, the cost per individual of providing relaxation training to relatively large numbers of employees is considerably less than in clinical programs since it is carried out in groups, takes little time overall (fifteen hours), and can be provided by skilled people who are not necessarily members of traditionally organized, recognized professions.

It is worth mentioning that since the time at which Schwartz (1982) drew attention to the absence of data enabling us to relate the effectiveness of specific stress management techniques to specific types of people under specified conditions, little evidence has been found of differential effectiveness (see Lehrer et al. 1983; Woolfolk et al. 1982). This affords some justification for training in a variety of techniques from which participants select according to their own multidimensional if unconscious matching criteria.

"Beyond Stress" was offered as either a two-day block course or as a five-week course. In both cases the total class time was about fifteen hours. In the case of the five-week program, however, there were homework assignments that clearly added to the practice time. Nevertheless, no differences of either statistical or practical significance were found between the outcomes of the two course formats, at least within the time frames considered in this report. Consequently, no further reference will be made here to the two formats. It should be noted, however, that Hillenberg and Collins (1983) found that home practice did provide superior benefits in their study.

The course exposed participants to a variety of relaxation techniques. Elements of transcendental meditation, yoga, and other more familiar somatic and autogenic methods were used. A listing of major components is shown below. Certain specific methods have been subject to evaluation among selected populations (for example, Orme-Johnson and Farrow 1978; Peters, Benson, and Porter 1977a, 1977b; Carrington et al 1980) and the "Beyond Stress" course was unusual in that participants had an opportunity to learn and practice a large number of methods known to be effective with at least some group or groups. By self-selection, it was hypothesized that participants would find those methods best suited to them. Consequently, the evaluation addresses this process of self-selection and its results.

The course was aimed at the learning of skills associated with the ability to relax. While there was also some emphasis on increased awareness of what stress is and how it manifests itself, most of the course was directed at the learning of various ways in which a state of relaxation can be generated *at will*. There was an emphasis on behavior in this course because participants were considered to have made a decision that they wanted to learn to relax and therefore did not need specific persuasion to attempt behavioral change. They can be said to have already formed behavioral intentions (Fishbein 1983).

However, this assumption was examined in the course of the evaluation to the extent that the auspices under which people came to the sessions were enquired about. For example, if coercion of any kind was involved (which appears to have been rare), this would clearly reduce the degree to which any assumption could be made about the desire to learn new behaviors. All the participants who remained in the research sample reported that they had joined the course willingly.

Major Components of the Course. The list that follows is adapted from the course outline provided to prospective participants by the instructor.

Dystension Exercises: A series of light exercises to loosen the body, integrate breathing and movement, eliminate tensions in the neck, shoulders, stomach, and back.

Abdominal Breathing: To correct improper breathing and establish deep, even, abdominal-centered breathing. The method shows how breathing can be brought under the mind's conscious control to enable one to quickly and effectively deal with stressful situations. The practicality of this method is emphasized and its daily application is encouraged.

Progressive Relaxation: By alternatively tensing and relaxing the skeletal muscles progressively through the body, deep states of relaxation can be produced. When the muscles are naturally relaxed, mental anxiety cannot be present.

Autogenic Training: Deep states of relaxation can be induced by imagining that parts of the body become very heavy and warm. Working through the parts of the body in a prescribed order can produce very relaxed states that are then deepened by various methods.

Serenity Breathing: An exercise in controlled breathing that slows down body metabolism and respiration and produces a state of mental and physical tranquillity.

Focusing the Mind: Techniques to focus the mind on counting and on verbal repetitions (such as "I am calm") to stop the ever-present "chatter" of the mind. When properly carried out this results in mental and physical relaxation, enhanced concentration, and higher energy levels.

Spiral Relaxation: A method to quickly relax the body by traversing it with the mind's eye.

Expanding Focus:	A technique that enables the letting go of narrowly focused attention through the practice of object-less imagery, for example, imagining the space between your eyes.
Visualization, Positive Affirmations:	Used in conjunction with deep relaxation to bring about a greater sense of balance between mind and body and to enhance the achievement of personal goals.
Differential Relaxation:	Learning to relax the body while moving so as to use only the required amount of muscle tension to accomplish your tasks.

Predicted Impact of the Course upon Alcohol and Drug Use

The hypotheses predicting the impact of the course upon alcohol and drug use were derived from a conceptual model in which the use of such substances was related to the process by which stress (external circumstances in the form of life events or conditions) becomes mediated through the coping repertoire of individuals to result in differing levels of strain (experienced or subjective distress, anxiety, and discomfort).

This more complex model of the stress, strain, coping process has emerged in the relatively recent past. Hitherto, the emphasis appears to have been upon a direct relationship between external events or stressors (life events) and health-related problems. However, in predicting who will experience subjective strain from exposure to potential stressors, the concept of coping has received an increasing amount of attention in recent years. Coping resources can be roughly divided into the areas of mental and physical health, morale, problem-solving skills, belief systems, social support, and material support (Roskies and Lazarus 1980). These may be further reclassified into three areas: personal characteristics or traits (for example, self-esteem, mastery, locus of control); personal skills (for example, interpersonal, problem-solving, cognitive); and personal resources (for example, social support, money, property). It is argued that more attention should be paid to these coping resources and how they are managed than to the characteristics of stressors themselves. That is, we are admonished to study the factors that mediate stress and either turn it to advantage, neutralize it, or convert it into subjective strain (Roskies and Lazarus 1980; Lumsden 1981). This formulation of the research task stands as a criticism of the life events approach in which direct relationships are sought between self-reported changes in one's life (divorce, death or illness of family member, promotions, demotions, job loss, etc.) and health status in both physical and mental domains. The life events approach to the study of stress and health has been most visibly associated with the work of Holmes and Rahe (1967) and later variations on their theme. The criticism—at least of the earlier work—was justified: correlations between the number of life changes in a given period and poor health status were positive, even statistically significant, but weak (Rahe 1975). The introduction of methods for assessing the individual's own *evaluation* of change events (Mules, Hague, and Dudley 1977, for example) and differentiation of desirable and undesirable events (Vinokur and Selzer 1973, for example) has led to some improvement in the ability of the method to predict future health status.

Currently, work in this area includes studies of the extent to which life events are perceived to be controllable and the degree to which the individual experiencing them feels responsible for their having occurred (Hammen and Mayol 1982; Suls and Mullen 1981; Sarason et al. 1983). Others have pointed to the need to consider chronic life conditions (socioeconomic status, poverty, long-term disability, illness, unemployment, overcrowding, isolation, etc.) as opposed to only life events in the study of experienced strain and distress (Makosky 1982).

However, these modifications are seen by some as attempts to improve an approach that is basically flawed in that it still emphasizes the stressor rather than the individual's way of coping with it. Lumsden (1981), for example, favors a transactional concept of stress in which the stimulus is only stressful (in the sense of over-taxing the individual's resources) if certain *meanings* are ascribed to it. These meanings are heavily influenced by cultural and social norms. For example, certain social groups may be more disposed to believe that they can exercise control over certain aspects of their environments than others. Thus, a sense of self-efficacy is socially taught and reinforced. Similarly, however, a sense of powerlessness and lack of control can be socially or culturally reinforced. In some cases, drinking—even to excess—is likely to be seen as one of the few available, acceptable coping responses to stresses. These are circumstances in which the prepotence of the reinforcement associated with drinking is likely to emerge (Zimering and Calhoun 1976).

It is evident from Cobb (1976) that social support can act as an important moderator of life stress. He reviews experimental and other evidence in which social support was shown to moderate the impact of such events as hospitalization, surgery, illness (tuberculosis, asthma, arthritis), depression, other psychological problems, job stress, death of a loved one, and birth complications. The way in which social support operates to produce these positive results is hypothesized to be through expression that one is loved, cared for, esteemed and valued, and belongs to a network of communication and mutual obligation (Cobb 1976, 300). Groen and Bastiaans (1975) note that there has been a disintegration of basic support networks in our society.

Smaller, more widely distributed families, rapid technological change, unstable communities, increased incidence of both parents working, anonymous forms of communication, and accessible transportation all create a situation in which people have to work at maintaining reliable support networks.

LaRocco, House, and French (1980) and Pinneau (1975; 1976) have raised questions about the way in which social support moderates job stress. While it appears that social support mediates perceptions of stress from some sources and not from others, it seems reasonable to hypothesize that social support, or the absence of it, is likely to be important in influencing the success of participants in stress management programs. (See chapter 9 for an elaboration of the role of social support in mental health and its practical implications.)

Within this model, substance use is considered to have a number of functions, not necessarily exclusive of one another. Furthermore, they are socially learned functions (Bandura 1977a; Pattison, Sobell, and Sobell 1977; Sobell and Sobell 1978), that is, individuals learn in the context of socially meaningful groups that alcohol and drugs will or will not do certain things for them under certain conditions. The primary functions of substance use that have been identified in the literature pertaining to the stress, strain, and coping process may be briefly summarized as follows.

1. The reduction of tension in physiological terms. This is one part of the tension reduction hypothesis reviewed by Cappell (1975).

2. The second part of the hypothesis is the raising of expectations that tension will be reduced. As a whole, the hypothesis suggests that alcohol (and by inference certain types of drugs) are either chemically effective in reducing tension, are believed to be, or both (Donovan and Marlatt 1980). Some postulate that interpersonal stress (discomfort with human interaction) may be a particularly strong stimulus for the use of alcohol, including excessive use (Higgins and Marlatt 1975; Marlatt, Kosturn, and Lang 1975). More evidence has been accumulated in support of the second part of the hypothesis than for the first.

3. The increase in experienced self-efficacy or mastery (Marlatt and Gordon 1980; Pearlin and Schooler 1978; Pearlin and Radabaugh 1976). This function may, as suggested, coexist with tension reduction functions. These concepts are generically related to aspects of alienation (Seeman 1959; Seeman and Anderson 1983) and locus of control research (for example, Johnson and Sarason 1978; Marlatt and Marques 1977). The drift of the theory underlying this postulated function is that alcohol acts upon the mind and body in such a way as to "empower" the individual.

The "coping" aspects of alcohol consumption may be more a function of the early phase of intoxication than of the later phase. Marlatt (1979) observes that the biphasic nature of alcohol's effects are such that while a lower dose may produce euphoric feelings (increased energy, excitement, sense of personal power or control), an increased dose may produce dysphoric effects. It is probably the immediate, euphoric effects that provide reinforcement for repeated episodes of drinking.

The present study, then, set out to address the following questions:

Does a popular relaxation course offered to a variety of employee groups achieve its stated objectives of relieving symptoms of distress and anxiety as a function of learning how to relax at will?

Does the course have any impact upon drinking behavior? Does the course have any impact upon drug use (prescription and over-the-counter) and coffee consumption? Does the course have any impact upon mastery (self-efficacy)? How does this effect relate to tension reduction and use of chemicals?

Method and Procedure

The pre- and post-design employed in this study was a variation on the quasi-experimental approach. The "Experimentals" (hereafter referred to as "Es") were employed people who volunteered to attend a "Beyond Stress" course designed and presented by the instructor. As already noted, the course was offered on either a two-day block basis in the employees' place of work or on a five-week, three-hours-per-week basis. In the five-week version, sessions were often held in the evenings at the offices of the instructor. In most cases the fees for the course were paid for by the employer.

The "Comparisons" (hereafter referred to as "Cs") were employed people either from the same workplace as the one in which the course was held or from another similar location. They were invited to participate in the completion of a stress appraisal survey, after which they were shown a film on the maintenance of healthy life styles both at baseline and first follow-up. Assurances of voluntariness and confidentiality were given verbally and in writing. In the case of both Es and Cs questionnaires were administered at five-week intervals. A third questionnaire was sent out six months after the second.

Survey Instruments. Questionnaires incorporating well-known instruments were designed for this study. The principal parts were as follows:

1. *The Life Experiences Survey* (Sarason, Johnson, and Siegel 1978). This is a measure of the degree of external stress placed upon the individual by virtue of changes to which adaptations have been required. It differentiates responses to life events in terms of their perceived positive or negative impact. Typical areas considered are marriage; acquisition of loans and property; promotions and demotions; divorce; change in living situation, eating habits,

sleep patterns; and illness or death of loved ones. The introduction of an evaluation component into the recording of life experiences impact was expected to produce better predictions of strain and distress (for example, high negative scores were expected to predict high distress levels more than high positive scores).

2. *Spielberger State-Trait Anxiety Inventory* (STAI). This inventory is one of the most widely used devices for the measurement of state and trait anxiety (Spielberger, Gorsuch, and Lushene 1970; Spielberger 1975). It has been employed in recent experimental studies of how alcohol consumption affects anxiety (Wilson, Abrams, and Lipscomb 1980; Bradlyn, Strickler, and Maxwell 1981; Hire 1978; Eddy 1979; Lipscomb et al 1980; Abrams and Wilson 1979; Logue et al. 1978). The STAI has also been used in a study of how group autogenic training affects both state and trait anxiety (Herbert and Gutman 1980). Herbert and Gutman found, rather surprisingly, that both state and trait anxiety levels were significantly decreased in comparison with a control group. State anxiety is seen as a *transitory emotional state* or condition of the human organism that is characterized by subjective, consciously perceived feelings of tension and apprehension and heightened autonomic nervous-system activity. Trait anxiety refers to a *relatively stable tendency* to respond to situations perceived as threatening with elevations in State Anxiety (Herbert and Gutman 1980, 112). The Beyond Stress course did cover autogenic training but not to the extent of the Herbert and Gutman experiment.

3. *Derogatis Symptom Checklist (SCL-90-R)* (Derogatis 1977). One of the main reasons for using this instrument is that it was used by Carrington and associates (1980) in a study of stress management techniques with a normal, employed population. This study is one of the best reported in the area so far, and it is advantageous to compare our population's profile on the SCL-90-R with that of Carrington's. The instrument is the latest version of the Hopkins Symptom Checklist (HSCL). Its history, validity, and development are reviewed by Derogatis, Lipman, and Covi (1973) and by Derogatis and Cleary (1977). Normative data for a variety of populations are available, making the device more useful. It covers nine areas, all or some of which may be affected by relaxation training. They are: somatization, obsessive-compulsive behavior, interpersonal sensitivity, depression, anxiety, hostility, phobic anxiety, paranoid ideation, and psychoticism.

4. *Pearlin Mastery Scale* (Pearlin and Schooler 1978). This is a conveniently short scale that has been used in several studies of depression, anxiety, and alienation and drinking (Pearlin and Radabaugh 1976; Ilfeld 1976; MacBride et al. 1981; Fields 1980). It is conceptually related to the notion of self-efficacy discussed earlier and as such is a measure of perceived control over the circumstances of one's life. Bandura (1977b) has proposed and to some extent demonstrated that individuals expect success (feel powerful)

when they are calm and rested. They expect failure (feel powerless and out of control) when they are agitated and upset. Consequently, we might predict that improvements in the ability to relax will bring about a greater sense of mastery. The process underlying this link may be that when people experience greater control over their own bodies and emotions through being able to relax at will, they generalize these feelings of control to at least some other areas of their lives.

5. *Frequency and Amount with Regard to Use of Alcohol, Prescription Drugs, Tobacco.* It is hypothesized that if participants in the Beyond Stress course report no use of these substances to begin with, they will maintain their behavior. If they use these substances already, it is hypothesized that they will either reduce consumption to safe levels (for example to no more than two drinks per day on the average) or discontinue use altogether (in the case of tobacco). In relation to drugs obtained on prescription, one would expect to see a reduction in the need for drugs that calm, elevate or stabilize mood, or bring sleep, since these are all anticipated effects of relaxation training. It is less clear what one should expect in the domain of drugs prescribed for other conditions. Self-prescribed drug use was also monitored, with particular attention paid to painkillers, vitamins, stimulants, and sleeping "aids." In this context we also enquired about the use of coffee and tea. Standard methods used by the Addiction Research Foundation's Clinical Institute were employed here (ARF, n.d.).

6. *Satisfaction with Eating, Sleeping, Exercise, Recreation, Work Habits.* An accumulating body of evidence lends weight to the argument that alcohol and drug abuse tend to develop in the context of life styles that are characterized by poor habits in eating, sleeping, exercise, recreation, and work (Miller 1980; Zook and Moore 1980; Wiley and Camacho 1980; Shain 1981). "Poor" in this context simply refers to the tendency for certain patterns of these behaviors—alone and in interaction—to be predictive of poor (worse than average for age–sex group) future health status.

Alcohol and drug abuse, and to some extent the use of tobacco, can be seen in the context of life style as a system of interdependent, interacting behaviors, beliefs, attitudes, and values. Consequently, it is appropriate to think of alcohol and drug abuse as being supported and reinforced by these other life-style system components. Interventions that affect these components are therefore relevant to the prevention of alcohol and drug abuse. In this project we enquired about personal *satisfaction* with habits in the areas indicated rather than attempting to construct a catalogue of behaviors. We expected to see some increase in dissatisfaction in the short run (between pre- and post-measures) and an increase in satisfaction as changes were introduced by the time of the six-month follow-up.

7. *Social Support.* This measure was developed in the well-known study by Brown, Bhrolchain, and Harris (1975), which suggested that supportive

social relationships could be protective against neurotic episodes among urban women particularly when faced with very trying circumstances. The Brown, Bhrolchain, and Harris measure of social support was used recently by MacBride and associates (1981) in a study of occupational stress among air transportation administration employees in Ontario. This opportunity for normative comparisons with a recent, local study seemed most useful. Inquiries about social support were made only at the second administration of the questionnaire.

8. *Tension Reduction Drinking* (Parker and Brody 1982). This scale was described in chapter 5. See particularly table 5–1, note e.

Hypotheses

Hypotheses in relation to the course were proposed in which, for the most part, the E group (those who were trained) were expected to become more "like" the C group as a result of exposure to the course. This approach clearly differs from a classic experimental design in which randomly allocated experimental and control groups are expected to diverge or become less alike on certain criterion variables as a result of some planned intervention aimed at the experimentals. The present design was adopted because it appeared impossible to obtain a "waiting list" control group owing to the immediate nature of the distress that led most people to apply for the course and to the unwillingness of sponsoring employers to comply with research requests from which they saw little chance of gain. In the design that was eventually adopted, therefore, the Cs were more of a normative or reference group in that as a whole their mean stress and strain scores were close to those reported by the originators of the instruments used in this study as being characteristic of "normal" populations. Thus, the primary approach to analysis that was eventually adopted was based upon the search for *convergence* between Es and Cs. The hypotheses, in reflecting these circumstances, were constructed to read as follows.

1. Initial differences between E and C groups in relation to their distress and mastery scores will be reduced significantly between first, second, and third administrations of the questionnaires, in that Es will become more like the Cs, whose scores will remain relatively stable.

2 a. Alcohol and drug consumption will co-vary positively with distress levels and negatively with mastery levels at baseline among both Es and Cs, that is, there will be more distress and less mastery among heavier users of alcohol and drugs.

2 b. Alcohol and drug consumption will decrease the most among those whose distress levels decrease the most and among those whose mastery levels increase the most. This will be most evident among Es.

2 c. Self-reported tension reduction drinking will diminish the most

among those whose distress levels are reduced the most and whose mastery levels increase the most. This effect will be most evident among Experimentals.

A secondary form of analysis was also conducted in which the Cs were split up into two groups according to their scores on three of the key measures discussed in the foregoing, namely the Spielberger State and Trait Anxiety scales and the Global Severity Index (GSI) of the Derogatis SCL-90-R. The procedure simply involved the designation of the lower scoring half ($n = 18$) of the Cs as the "low normals" and of the higher scoring half as the "high normals." In the case of all three measures, this procedure resulted in two groups with significantly different mean scores at baseline (before the course). The scores of the high normals were found to approximate those of the Es, thus providing a population that would have been referred to as a control group if the allocation of subjects to it had been random. This form of analysis was undertaken in order to provide an alternate test of the argument that the observed changes in the E group were due to the course and not to chance. However, the results of both primary and secondary forms of analysis should be weighed together since they are complementary.

Results: Characteristics of the Sample

Sixty-three Es and thirty-eight Cs completed questionnaires on a before-and-after basis. A considerable proportion of the original 170 Es did not respond to the second questionnaire. Most of these people had been participants in the two-day workshops. Motivation to comply with research requests five weeks after the course was low. Overall, the pre–post match rate for Es was 37 percent while for Cs it was 70 percent. The higher rate among Cs was probably due to the fact that in return for completing the questionnaires a second time they were shown another film on healthy life styles during work hours.

Research Completers vs. Research Drop-Outs. There does not appear to have been a great deal of difference, however, between those who completed the second administration and those who did not, at least according to the variables measured by the pre-course questionnaire. In age, sex, education, income, and marital status there were no significant differences, nor in scores related to anxiety and distress. Alcohol consumption was about the same in both groups. In so far as we can detect trends at all, it seems that the research drop-outs contained a greater proportion than did the completers of people who were less troubled by work, health, medication use, and the amount of sleep they were getting. Prior to the course, they had been somewhat less interested in self-help. They tended to smoke less than the research completers. While this evidence does not allow us to safely generalize from the 63 research completers to the other 107 Es who completed only the first administration of the questionnaires, at least it does not exclude the likelihood that the findings reported here are of relevance to a considerable proportion of the research drop-outs.

Experimentals vs. Comparisons. Both groups contained a greater proportion of relatively well-educated, high-income people doing mainly white-collar jobs than might be expected in the general population or in the work force as a whole. Males and females were equally represented in both groups. Both groups were still fairly young, the mean age being considerably below forty. The majority of the respondents in both groups were married or in some other stable relationship, although the separation and divorce rate for Cs was greater than among Es. Even so, Cs were more likely to have had children than Es, although family size did not differ among those in both groups who had children. Es were somewhat more likely to have been born outside Canada.

The two groups varied to some extent in their desire to have changes in their work situations and in their eating and sleeping habits. Such differences may be reflected in the significant variation between groups relative to anxiety and distress, and in self-reported difficulties with relaxation, in spite of past efforts to overcome them. These differences will be reviewed in the presentation of the evaluation results.

Little difference was found in terms of medication, alcohol, and tobacco use. No difference at all was found between groups on tension reduction drinking and mastery scores.

On the whole, the use of the C group as a referent for changes that might have been expected to occur among the Es seems to have been warranted.

Analysis. Given the nature of the E and C groups in this study, the analysis of course impact was predicated on the assumption that initial differences between the groups on key variables relating to distress would tend to diminish if the intervention was successful. Thus, a principal form of analysis was the comparison of significant differences between groups before and after the course. This was achieved by separate analyses of variance (ANOVAs) on baseline and follow-up data. If significance levels decreased this was taken as support for the hypotheses relating to the distress and mastery variables. For the study of within group changes over time we employed T tests. A reading of the two sets of results, ANOVAs and T tests, provided a fairly complete picture of what had occurred and whether it was of statistical significance. However, such analyses in themselves provide only circumstantial evidence of the reasons for such changes and do not rule out the influence of factors other than the course.

Within the framework of the analytic approach just outlined, two methods were used in order to strengthen the conclusions that could be drawn from the available data. First, we examined the comprehensiveness and interrelatedness of observed changes by performing correlations among them. Patterned changes, we believed, would be more indicative of course-related effects than isolated changes. Second, for purposes of examining the second hypothesis relating to alcohol and drug use we adopted an approach that will be referred to hereafter as *risk group analysis*.

This was an attempt to control for pre-test variance between Es and Cs by constructing *risk groups* based on their scores derived from a quantity–frequency (Q–F) index of alcohol consumption. In this manner, groups with equivalent baseline mean scores on the Q–F index were constructed for both Es and Cs. Thus, it was possible to compare the changes occurring within two high-risk, two moderate risk, and two low-risk groups, in each case one from the Es being matched with one from the Cs. In this approach, however, the proportion of Cs in the high-risk category was smaller than in the case of Es.

Results on Hypothesis 1

This hypothesis was tested in relation to change scores on stress and strain variables and mastery.

Spielberger State and Trait Anxiety. There was a significant difference between male E and C groups at baseline (pre-test) in relation to state anxiety (ANOVA $p > .001$). By the end of the course or after five weeks (in the case of the two-day version) this difference had disappeared ($p > .50$). Among females, a nearly significant difference ($p > .07$) also disappeared ($p > .22$).

Although the difference between E and C groups was not significant at baseline in relation to trait anxiety ($p > .27$ for males and $p > .21$ for females) these differences nonetheless became even smaller after five weeks (males, $p > .90$; females, $p > .41$). The actual amount of change is shown on table 6–1.

It can be seen from table 6–1 also that the ANOVA changes are a reflection of significant within-group changes among Es that do not occur among Cs. This is confirmed in T tests.

Changes can also be evaluated in the context of the norms provided in the test manual (Spielberger, Gorsuch, and Lushene 1970). Among E males, mean baseline scores on state anxiety were comparable to those obtaining at the 45th percentile of the score distribution among male neuropsychiatric patients. After five weeks, the mean state anxiety scores of E males were comparable to those found at the 28th percentile of scores among neuropsychiatric patients. In the meantime, C group mean scores for males had gone from those obtaining at the 18th percentile for neuropsychiatric patients

Table 6–1
Pre- and Post-test Scores: Anxiety and Distress

	Males		Females	
	E	C	E	C
	(n = 36)	(n = 17)	(n = 27)	(n = 21)
Spielberger State Anxiety Mean Scores[a]				
Pre	47	35	46	40
Post	39	37	39	43
Spielberger Trait Anxiety Mean Scores[b]				
Pre	40	37	42	38
Post	36	36	37	40
Derogatis SCL-90-R GSI Raw Scores[c]				
Pre	0.54	0.29	0.87	0.54
Post	0.31	0.30	0.55	0.49

[a]T tests (pre to post, sexes combined): Experimentals, T = 5.62, P > .001; Comparisons, T = −0.31, P > .76.
[b]T tests (pre to post, sexes combined): Experimentals, T = 6.74, P > .001; Comparisons, T = 0.85, P > .40.
[c]T tests (pre to post, sexes combined): Experimentals, T = 7.48, P > .001; Comparisons, T = 0.65, P > .52.

to those obtaining at the 23rd. For the E males, the change can also be expressed as movement from mean scores obtaining at the 87th percentile for undergraduates to the 65th.

Among females, the neuropsychiatric norm is not available. The change among E females, therefore, is expressed as movement from mean scores obtaining at the 87th percentile for female undergraduates to the 71st. This can be seen in the context of C females who went from the 74th to the 83rd percentile for undergraduates. Table 6–2 provides further context for interpretation of the scores and changes seen in our sample. Relative to the baseline scores in other studies, it appears that our sample tended to score somewhat higher on state anxiety and somewhat lower on trait anxiety. The magnitude of changes seen in our sample, however, seems to be well within the bounds of changes seen in the other studies despite variations in follow-up periods. It is safe to say that, based on all the evidence so far, the observed reduction in state and trait anxiety in the present sample is of practical, clinical importance.

Derogatis SCL-90-R Global Severity Index (GSI). Baseline significant differences between E and C groups (ANOVA p > .02 for males, p > .05 for

Table 6–2
Comparison of Present Sample with Samples in Other Studies on the Spielberger Scales

Study/Treatments		Baseline	Spielberger Scores (Change) (Δ)	Follow-Up[a]
Friedman, Lehrer, and Stevens (1983)				
(General Stress Management with	State A	44	(−10)	34
Teachers)	Trait A	41	(−7)	34
Lehrer et al. (1983) (High Scoring Volunteers and Referrals)				
Progressive Relaxation	State A	42	(−5)	37
	Trait A	52	(−7)	45
Meditation	State A	43	(−4)	39
	Trait A	54	(−4)	50
Waiting List Control	State A	33	(+6)	39
	Trait A	51	(−1)	50
Hillenberg and Collins (1983) (Volunteers)				
Progressive Relaxation (with homework)	Trait A	49	(−7)	42
P.R. (no homework)	Trait A	51	(−5)	46
Waiting List Control	Trait A	50	(+1)	51
Present Sample (Shain and Bay, 1985)		M/F	M/F	M/F
Relaxation	State A	47/46	(−8/−7)	39/39
	Trait A	40/42	(−4/−5)	36/37
Comparison	State A	35/40	(+2/+3)	37/43
	Trait A	37/38	(−1/+2)	36/40

[a]Varies from 7 weeks to 6 months.

females) disappeared by the time of the post-test ($p > .93$ for males, $p > .67$ for females). This corresponds to a change among E males whereby at baseline their mean score was characteristic of the 88th percentile of scores in the norm group and at follow-up it was characteristic of the 70th percentile. Among C males, the corresponding change was a slight increase (as in the case of Spielberger State Anxiety).

Among females the mean score for Es went from a point characteristic of the 90th percentile of scores in the norm group to one characteristic of the 80th. Among C females there was a decrease from the 80th to the 75th percentile scores found in the norm group. Otherwise stated, in the case of both males and females, GSI scores went from those in the "subclinical" range

Table 6–3

Comparison of Present Sample with Samples in Other Studies on the Derogatis SCL-90-R Global Severity Index

| Study | GSI Scores | | |
	Baseline	(Change)	Follow-Up
Carrington et al. (1980)			
Experimentals	0.74	(−.44)	0.30
Controls	0.80	(−.33)	0.47
Present Sample			
(Shain and Bay 1985)	*M/F*	*M/F*	*M/F*
Experimentals	.54/.87	(−.23/−.32)	.31/.55
Comparisons	.29/.54	(+.10/−.05)	.30/.49

(2nd standard deviation) to those in the "normal" range (1st standard deviation). In the foregoing discussion, the norm group referred to is a randomly selected sample of citizens (n = 494 males, 480 females) living in a representative county of a large U. S. eastern state (Derogatis 1977).

The baseline GSI scores in our sample are quite similar to those found by Carrington and associates (1980) in their study of New York Telephone Co. employees who self-recruited themselves into a stress management program run by that organization. The similarities become more apparent when a rough average of male and female scores in our sample is taken (table 6–3). Carrington and associates did not report sex-specific data. As in the case of the Spielberger scores, it is clear that the reduction in distress signified by the drop in GSI scores is of practical, clinical importance.

Relationship of Positive to Negative Stressors. As noted earlier, stress on the individual (as opposed to the experience of strain from this stress) was measured by the frequency of life events requiring some sort of adaptation. These events could be evaluated subjectively in either a positive, negative, or neutral manner, but in all cases they were considered as changes. The time referent at pre-test was the preceding year and at post-test it was the previous five weeks. For this reason the two estimates are not directly comparable but pre-test scores for Es can be compared with pre-test scores for Cs while a similar operation can be performed at post-test. The results are displayed in table 6–4.

It can be seen that Es and Cs were quite similar in their combined positive and negative scores. However, among the Es we find that whereas at baseline negative events were reported more frequently than positive events, the re-

Table 6–4
Life Experiences Survey[a] Scores: Negative to Positive Relationships, Pre
to Post

	E		C	
	Pre-Test (Last Yr.)	*Post-Test (Last 5 weeks)*	*Pre-Test (Last Yr.)*	*Post-Test (Last 5 weeks)*
Total Positive	4.2	3.7	5.4	2.7
Total Negative	6.9	2.3	6.1	2.7
Combined ±	11.1	6.0	11.6	5.4
Net Δ ±	− 2.7	+ 1.4	− 0.7	0.0

[a]Sarason, Johnson, and Siegel (1978).

verse occurs at follow-up. Among Cs negative events barely outstripped positive events at baseline, while at follow-up they matched one another. The relationship between positive and negative events can be expressed as a difference score derived by subtracting negative from positive. These results are described in table 6–4 as Net Δ ±. Superficially, they seem to indicate that Es either learned to reinterpret changes that had already occurred as positive rather than negative or, more simply, that the course was seen as engendering a number of positive events.

One might expect that negative life events would be more related to anxiety and distress than positive events. This indeed appears to be the case (see table 6–5).

However, among Es one notes that a small but significant correlation between positive events and GSI developed between the two questionnaire administrations. Among Cs no such relationship was found at either administration (table 6–5). The same trend is not observed between positive events and state or trait anxiety.

While it must be remembered that Es decreased significantly on all key distress scores, the observations described here seem to suggest that if life events are seen as representing *changes to which adaptations are required,* then the E group appears less able than Cs to differentiate positively evaluated change from negatively evaluated change in terms of how it affects their distress levels.

This may suggest a way in which the self-selected Es are different from the Cs: namely, that the former tend to react in a distressed manner to *any* change, positive or negative, while the latter react primarily to negatively evaluated change. However, an alternate explanation of the slight increase in the relationship between positive events and distress among Es is as follows. As mentioned earlier, the course in relaxation to which Es were exposed may

Table 6–5
Pearson Product-Moment Correlations between Life Events,[a] Distress,[b] and Anxiety[c] before and after the Course

	E		C	
	Time 1	Time 2	Time 1	Time 2
Positive Life Events and GSI	0.17	0.23[d]	0.04	0.03
Positive Life Events and State Anxiety	0.11	−0.16	−0.13	−0.11
Positive Life Events and Trait Anxiety	0.11	0.02	0.06	−0.04
Negative Life Events and GSI	0.51[e]	0.50[e]	0.30[d]	0.64[e]
Negative Life Events and State Anxiety	0.40[e]	0.27[d]	0.20	0.27[d]
Negative Life Events and Trait Anxiety	0.13	0.27[d]	0.41[d]	0.54[e]

[a]Sarason, Johnson, and Siegel (1978).
[b]Derogatis (1977).
[c]Spielberger, Gorsuch, and Lushene (1970).
[d]P between > .05 and .01.
[e]P > .001.

have led them to reevaluate certain changes in their lives. For example, seen from a fresh (relaxed) perspective, an event that was previously evaluated as negative in its impact might be reevaluated as positive. Divorce, separation, and job promotion might be examples of this. However, the cognitive reappraisal of these events as positive may not always be synchronized with emotional reappraisal: hence at least a short-term dissonance between cognition and emotion. Whether this reappraisal actually takes place and the manner in which this dissonance is resolved are questions that cannot be answered by reference to the data at hand.

Mastery. It can be seen from table 6–6 that no significant difference existed *between* E and C at either administration 1 or 2, although, if anything, the difference grew smaller. Nonetheless, change took place within both groups, according to T tests. The change among Es was of greater magnitude but in both cases it was significant. Consequently, we cannot attribute the changes seen among Es to the course given this evidence alone. However, among male Es the observed changes in mastery were correlated with changes in Positive Symptom Total ($r = -0.27$; P > .01); ability to relax ($r = 0.39$, P > .01); State Anxiety ($r = -0.57$; P > .001) and Trait Anxiety ($r = -0.45$; P > .003). These correlations were not seen among Cs, nor among female Es, although in the latter case changes in mastery and trait anxiety were related to some extent ($r = -.29$; P > .08). See figure 6-1 for details.

It appears, then, that for males at least there was an important pay-off

Table 6–6
Changes in Mastery[a]

	E	C
Pre	22.0	22.7
Post	23.2	23.6
ANOVA (oneway)	Admin. 1: F = 0.777; P > .38	
	Admin. 2: F = 0.250; P > .62	
T tests (Pre to Post)	Experimentals T = −2.88 P > .006	
	S.D. of difference = 3.29	
	Comparisons T = −2.49 P > .02	
	S.D. of difference = 2.24	

[a]Pearlin and Schooler (1978).

in feelings of improved mastery or self-efficacy when relaxation training was successful.

Before the course, substantial correlations were found between mastery and mental health measures in both E and C groups. Thus, it cannot be argued that the correlations between improvements in these dimensions seen in the Es were a function of greater initial relationships between the same variables prior to the intervention. A table of the correlations seen at baseline follows (table 6–7).

Fate of Hypothesis 1

It appears that the first hypothesis is largely supported by the data with regard to alleviation of distress among the Es and is partially supported with regard to mastery.

Results of Secondary Analysis. As noted earlier, the C group was split into "low and high normals" based on their pre-course scores on three mental health measures. The high normal group scores closely approximated those of the Es, thus providing a basis for an alternate examination of the importance of changes seen among the latter between the two questionnaire administrations.

The results indicate that the high normal Cs did not undergo significant changes, in contrast with Es. This evidence weighs heavily against the view that Es would have improved "spontaneously" without the relaxation course since, were that so, one would have seen such an effect also among high normal Cs (table 6–8).

Table 6–7
Pre-Course Pearson Product-Moment Correlations between Mastery and Mental Health Measures

| | Experimentals | | | | Comparisons | | | |
| | Males (n = 34,36) | | Females (n = 24–27) | | Males (n = 16,17) | | Females (n = 20,21) | |
Variable	R	Sig.	R	Sig.	R	Sig.	R	Sig.
Total (+) Life Events[a]	—		—		—		—	
Total (−) Life Events[a]	—		—		−.42(.046)		−.68(.000)	
State Anxiety[b]	−.56(.000)		—		−.51(.022)		—	
Trait Anxiety[b]	−.73(.000)		−.54(.003)		−.59(.008)		−.46(.021)	
Somatization[c]	—		−.36(.032)		—		−.44(.024)	
Obsessive-Compulsive[c]	−.43(.005)		(0.26)(.099)		—		−.63(.001)	
Interpersonal Sensitivity[c]	−.58(.000)		−.34(.040)		−.46(.30)		−.53(.006)	
Depression[c]	−.48(.001)		(−.30)(.061)		−.51(.018)		−.69(.000)	
Anxiety[c]	−.42(.005)		—		−.58(.007)		—	
Hostility[c]	−.55(.000)		—		−.55(.011)		—	
Phobic Anxiety[c]	−.46(.002)		—		—		—	
Paranoid Ideation[c]	−.57(.000)		−.37(.030)		−.75(.000)		−.46(.019)	
Psychoticism[c]	−.43(.004)		−.35(.038)		—		−.45(.021)	
Positive Symptom Total[c]	−.65(.000)		(−.28)(.077)		−.58(.007)		−.57(.004)	
GSI[c]	−.58(.000)		−.35(.036)		−.51(.019)		−.54(.005)	
Tension Reduction Drinking	—		—		—		−.56(.005)	
Painkiller Use	—		(.29)(.075)		—		(+.30)(.096)	

Note: Correlations with significance levels less than .05 are given in parentheses.
[a]Sarason, Johnson, and Siegel (1978).
[b]Spielberger, Gorsuch, Lushene (1970).
[c]Subscales and total scores from the Derogatis SCL-90-R (Derogatis, Lipman, and Covi 1973).

Results on Hypothesis 2

Drug Use. The reported frequency of prescription and over-the-counter drug use in the previous month is shown in table 6–9. The rate of use is well within the bounds of estimates recently assembled from the literature by Shain, Boutilier, and Simon (1984). One might have expected a group particularly troubled by stress to have used more tranquillizers than the population at large but this appears not to be the case. Indeed, Cs reported more use of such drugs than Es. Almost all the calming medications that respondents referred

Table 6–8
Pre-Post Course Changes on Key Mental Health Measures When Comparisons Are Split into High and Low Normals

Questionnaire Administration #	Experimentals	Scores (Sexes Combined)	
		High Normal Comparisons	Low Normal Comparisons
		State Anxiety	
1	46	45	31
2	39	43	34
		T = 2.00	T = −1.62
		P < .06	P < .12
		(*n* = 18)	(*n* = 18)
		Trait Anxiety	
1	41	45	30
2	36	43	31
		T = 1.56	T = −1.04
		P < .14	P < .31
		(*n* = 18)	(*n* = 18)
		Global Severity Index (SCL-90-R)	
1	.68	.68	.18
2	.41	.63	.18
		T = .67	T = .02
		P < .51	P < .98

to by name were found to be prescribed when checked against the Compendium of Pharmaceuticals and Specialties (CPS 1984).

In the case of sleeping medication the situation was somewhat different. The high rate of use of these drugs in both E and C groups conceals the fact that only just over half the named drugs were prescription. Even so, some respondents were clearly using other drugs—for example, nonprescription painkillers—to help them sleep. For purposes of some analyses the reported function of the medication was allowed to stand.

Although the number of drug users among Es and Cs separately was too small to warrant pre and post comparisons, except in the case of painkillers, it was still possible to consider the baseline characteristics of such users. It was found that use among Es and Cs at baseline appeared to be correlated with the same variables irrespective of whether they had completed the research follow-up or not. Consequently, we combined Es and Cs, drop-outs

Table 6–9
Prescription and Over-the-Counter Drug Use before and after the Course

Prescription and Over-the-Counter	Pre		Post	
	E	C	E	C
Sleeping Medications	15.9%	13.2%	11.1%	13.2%
Pain Medications	46.8%	57.9%	36.5%	63.2%
Calming Medications	4.8%	7.9%	4.8%	5.3%

Prescription Only (E and C groups combined)[a]

Sleeping Medication 5.5% ($n = 238$)
(an additional 2.9% were cited as prescription drugs but were not named)
Calming Medication 4.2% ($n = 241$)
(an additional 1.65% were cited as prescription drugs but were not named)

Sleeping Drugs Listed (checked with CPS)

Halcion (Triazolam) hypnotic   Somnol (flurazepam) hypnotic
Vivol (Diazepam) sedative/anxiolytic   Zapex (oxazepam) anxiolytic
Mandrax (methaqualone) hypnotic   Tuinal (secobarbital sodium) hypnotic
Ludiomil (maprotiline) antidepressant

Calming Drugs Listed (checked with CPS)

Serax (oxazepam) anxiolytic   Butisol (Sodium butabarbital) daytime sedative/hypnotic
Visken—antihypertensive   Lorazepam (benzodiazepine) anxiolytic
Trifluoperazine—antipsychotic
Librax (Chlordiazepoxide HCl)

[a]Only those drugs specifically referred to by name and then found to be available only on prescription (by reference to the CPS) were counted under this heading.

and completers alike, in order to provide sufficient numbers for analysis of variance between users and nonusers on a series of key variables. The results of these comparisons are shown on table 6–10.

Calming Medications. Users were significantly higher on most strain and distress measures than nonusers and were lower on mastery. It is interesting, however, that users differed from nonusers in relation to *positive* stressor (life event) scores but not *negative* stressor scores. Users had higher positive event scores than nonusers. This is different from the situation among users of sleeping medications who scored higher on negative events than nonusers. Intuitively this suggests that calming-drug users tend to be just as disturbed about relieving the strain that comes from coping with the "good" things of life as they are about the "bad," and that sleeping-drug users tend to be more

Table 6–10
Drug Use and Key Variables at Baseline[a]

Variables	Scores		ANOVA	
	Users	Nonusers	F	Sig.
Calming Medication: Characteristics of Users and Nonusers				
State Anxiety	49	45	1.072	n.s.
Trait Anxiety	48	40	8.665	.004
Positive Life Events	8.1	4.6	6.102	.01
Negative Life Events	6.6	7.3	0.092	n.s.
GSI (Derogatis)	1.09	0.60	14.543	.002
PST (Derogatis)	50	32	11.544	.001
Mastery	19.9	22.5	5.873	.02
Alcohol Q–F Index	5.8	8.1	1.665	n.s.
Tension Reduction Drinking	12.2	12.9	0.921	n.s.
	$n = 14$	$n = 223$		
Sleeping Medication: Characteristics of Users and Nonusers				
State Anxiety (Spielberger)	49	45	2.655	0.10
Trait Anxiety (Spielberger)	43	40	1.318	n.s.
Positive Life Events	5.4	4.7	0.308	n.s.
Negative Life Events	10.7	6.9	5.549	0.02
GSI (Derogatis)	0.78	0.61	2.796	0.10
PST (Derogatis)	36	33	0.729	n.s.
Mastery	21.3	22.5	2.158	n.s.
Alcohol Q–F Index	6.4	8.1	1.554	n.s.
Tension Reduction Drinking	13.6	12.8	1.438	n.s.
	$n = 23$	$n = 215$		

concerned about redressing irregular rest patterns disrupted by the impact of unpleasant events or circumstances. Users of calming drugs see themselves as having less control over their lives (lower mastery), given that their anxiety levels tend to be at the mercy of any change in circumstances, good or bad, to which they must adapt.

Sleeping Medication. As noted above, users of sleeping drugs tended to have much higher negative life event scores than nonusers but they appear to be quite similar to nonusers with regard to the other measures of distress. Near significant differences are seen in state anxiety and the GSI of the SCL-90-R but nowhere else.

Painkillers. Users of these drugs tended to have significantly higher scores than nonusers on negative life events and on state anxiety.

Table 6–10 continued

Painkiller Medication: Characteristics of Frequent, Infrequent, and Nonusers

Variables	Scores			ANOVA	
	Freq. Users	Low Use	Nonuse	F	Sig.
State Anxiety (Spielberger)	49	43	45	3.102	.05
Trait Anxiety (Spielberger)	42	39	40	1.605	n.s.
Positive Life Events	4.6	4.6	5.9	0.127	n.s.
Negative Life Events	9.0	8.0	5.8	3.869	.02
GSI (Derogatis)	0.73	0.56	0.61	1.754	n.s.
PST (Derogatis)	38	31	32	2.313	.10
Mastery	22.9	22.2	22.3	0.514	n.s.
Alcohol Q–F Index	9.5	7.4	7.7	1.648	n.s.
Tension Reduction Drinking	12.6	13.0	12.9	0.325	n.s.
	$n = 49$	$n = 77$	$n = 108$		

[a]Experimentals and Comparisons combined; research completers and drop-outs combined.

Because the absolute numbers of cases were limited in the case of sleeping and calming drugs we decided to drop them from analyses of changes and to concentrate instead on the use of pain medication where the sample size was adequate for valid pre/post comparisons.

Of the 47 percent among Es who reported painkiller use in the month prior to the course, about one-fifth had obtained them on prescription. The frequency of use of these substances was converted into a metric score that enabled us to make pre and post comparisons. The outcome is shown on table 6–11. Clearly, Es reduced their consumption significantly more than Cs among whom there was an increase, largely attributable to males. The reduction in consumption, however, appears to have been primarily among those who used painkillers on between four and seven days during the month rather than among the heaviest users (table 6–12). In all, 32 percent of the Es decreased their use of painkillers while 58 percent remained the same (including nonusers) and 10 percent increased.

Reduction in the use of painkillers between first and second questionnaire administrations was strongly associated with reductions in trait anxiety ($R = 0.45$) and state anxiety ($R = 0.41$). In addition, reduced intake of painkillers was related to reduced intake of alcohol ($R = 0.37$).

Coffee. There was a moderate reduction in the use of coffee among E males from 4.1 cups per diem to 3.6 ($T = 2.84$, $p > .007$). No decrease was seen among Cs whose baseline levels of consumption were almost identical to those of Es. Of the 34 percent who decreased their consumption, over three-

Table 6–11
Use of Painkillers before and after the Course: Frequency of Use Scores

	Experimentals		Comparisons	
	Males	*Females*	*Males*	*Females*
Pre	0.69	1.85	0.80	1.76
Post	0.31	1.23	1.20	1.66
	T = 2.07	T = 1.96	T = −1.28	T = 0.17
	P > .05	P > .06	P > .23	P >. 86

Note: ANOVA (one-way, sexes combined) between E and C: Pre-test F = 0.543; P > .46; Post-test F = 6.163; P > .01.

Table 6–12
Use of Painkillers before and after Course: Actual Use

	Days Used in Last Month						
	None	*1 day*	*2–3 days*	*4–7 days*	*8–12 days*	*>12 days*	*n*
E after	40 (63%)	10 (16%)	8 (13%)	1 (2%)	2 (4%)	2 (4%)	63
E before	33 (53%)	6 (10%)	9 (14%)	10 (26%)	1 (2%)	3 (6%)	62
C after	14 (37%)	7 (18%)	9 (24%)	3 (8%)	3 (8%)	2 (5%)	38
C before	16 (42%)	7 (18%)	9 (24%)	0	4 (10%)	2 (5%)	38

quarters did so by one cup per day, while the remainder cut down by two or more. Fifty-six percent did not change while 10 percent increased their consumption.

Alcohol Use: Quantity–Frequency (Q–F) Index. This index was constructed by multiplying frequency of drinking occasions by number of drinks per typical drinking occasion. Frequency was recoded so that 0 = no drinking occasions during the last month and 6 = every day. Number of drinks per typical occasion was entered in its raw form. For example, someone who drank every day and usually had two standard drinks would score 12 on this index. The same person taking only one drink would score 6. The mean scores, before and after, were:

Es Before 8.4 (S.D. 6.13)

Cs Before 6.9 (S.D. 6.32)

$$F = 1.274 \quad p > .26$$

Es After 7.3 (S.D. 5.80)

Cs After 6.8 (S.D. 6.08)

   F = 0.166   p > .68

Thus, the difference in means between the two groups, never very important, diminished between the two questionnaire administrations.

Attention to the group means alone, however, is not adequate. As noted earlier, risk group analysis was employed to determine whether any effect could be found when baseline consumption levels (Q–F index scores) were controlled. Thus, three groups were constructed for males and two for females since there were not enough women in the heaviest drinking category to warrant separate attention. The resultant groups were:

	Experimentals—Sample Size	
	Male	*Female*
Q–F 1 (Light drinkers)	13	16
Q–F 2 (Moderate drinkers)	12	{10
Q–F 3 (Heavier drinkers)	9	

	Comparisons—Sample Size	
	Male	*Female*
Q–F 1 (Light)	5	18
Q–F 2 (Moderate)	5	{3
Q–F 3 (Heavier)	7	

Differential course effects were observed according to Q–F group status at baseline (table 6–13). Among E males the impact is largely one of reinforcing already moderate drinking practices. This reinforcement effect appears again in relation to tension reduction drinking where we also saw a significant reduction in the use of alcohol to relieve tension or to change mood among male moderate drinkers.

Among E females there was a drop in consumption among the combined moderate heavy drinker group but it did not reach the point of statistical significance. Even so, it is important to note that tension reduction drinking scores dropped significantly in this group.

Among Es we found also that male moderate drinkers (Q–F group 2) experienced some of the greatest positive changes on the Spielberger anxiety measures and the Derogatis global scores. These data are shown on table 6–14. This table also illustrates the rather unexpected finding that at baseline higher consumption levels were not a characteristic of respondents with

Table 6–13
Alcohol Quantity–Frequency Index Changes (Statistical Summary)

	Pre	Post	2 Tailed P.
Male Experimentals			
Q–F Group 1	3.5	4.0	n.s.
Q–F Group 2	8.7	6.3	T = 2.31 P > .04
Q–F Group 3	16.0	14.7	n.s.
Male Comparisons			
Q–F Group 1	4.0	2.2	n.s.
Q–F Group 2	8.4	8.2	n.s.
Q–F Group 3	15.9	14.3	n.s.
Female Experimentals			
Q–F Group 1	3.6	4.4	n.s.
Q–F Groups 2/3 (combined)	14.3	11.3	n.s.
Female Comparisons			
Q–F Group 1	2.9	3.7	n.s.
Q–F Groups 2/3	12.6	13.6	n.s.
Tension Reduction Drinking Score[a] by Q–F Group			
Male Experimentals			
Q–F Group 1	13.7	13.6	n.s.
Q–F Group 2	11.9	13.3	T = 2.29 P > .04
Q–F Group 3	11.0	10.8	n.s.
Male Comparisons			
Q–F Group 1	15.6	15.5	n.s.
Q–F Group 2	12.0	13.4	n.s.
Q–F Group 3	12.3	13.0	n.s.
Female Experimentals			
Q–F Group 1	14.1	13.8	n.s.
Q–F Groups 2/3	10.2	12.1	T = −4.30 P > .002
Female Comparisons			
Q–F Group 1	14.6	14.8	n.s.
Q–F Groups 2/3	13.0	12.7	n.s.

[a]Higher tension reduction scores indicate less tension reduction drinking. Scale derived from Parker and Brody (1982).

Table 6–14
Differential Impact of Course according to Quantity–Frequency Index of Alcohol Use (Experimentals Only)

	Pre-course	Post-course	T	Sig.(Pre–Post)
Spielberger State Anxiety Scores				
Males				
Q–F 1	49	45	1.55	0.14
Q–F 2	45	34	2.85	0.02
Q–F 3	49	36	3.05	0.02
Females				
Q–F 1	46	40	3.40	0.005
Q–F 2/3	47	35	2.27	0.05
Spielberger Trait Anxiety Scores				
Males				
Q–F 1	44	41	2.65	0.02
Q–F 2	38	31	3.75	0.003
Q–F 3	39	36	1.50	0.17
Females				
Q–F 1	43	37	3.52	0.004
Q–F 2/3	41	36	3.60	0.006
Derogatis Positive Symptom Total (SCL-90-R)				
Males				
Q–F 1	34	25	3.90	0.002
Q–F 2	31	15	3.92	0.002
Q–F 3	27	23	1.39	0.20
Females				
Q–F 1	42	29	5.78	0.001
Q–F 2/3	40	30	2.83	0.02
Derogatis Global Severity Index (SCL-90-R)				
Males				
Q–F 1	0.63	0.43	3.78	0.003
Q–F 2	0.56	0.21	3.15	0.009
Q–F 3	0.47	0.30	3.52	0.008
Females				
Q–F 1	0.93	0.59	4.19	0.001
Q–F 2/3	0.83	0.51	3.13	0.01

higher anxiety and distress levels. In fact, if anything the reverse was true: light drinkers were higher than either moderate or heavier drinkers on trait anxiety, positive symptom total, and GSI. Even so, the greatest impact of the course in terms of symptom alleviation was not upon light drinkers but upon moderate drinkers followed by heavier drinkers.

It is likely, however, that certain people with very high anxiety levels at baseline were not drinking or were drinking less because they were taking medication for their condition. Analysis of Variance does show that the use of sleeping drugs, calming drugs, and painkillers are all related to higher anxiety levels albeit in different ways. Further, when drug use is broken down according to Q–F Score we find that those taking prescription painkillers had lower Q–F scores. (Prescription painkiller users' Q–F score, 6.30; nonprescription users, 8.93. ANOVA, oneway, $F = 3.874$, $P > .05$.) As in the case of alcohol use and measures of distress, the predicted relationship between alcohol use and mastery did not hold. At baseline, lower mastery was a characteristic of those who consumed less alcohol among both E males and females and among C females but not males (table 6–15).

Among E males, there appears to have been a significant increase in mastery among heavier consumers and some increase among moderate consumers. Among E females the opposite trend prevailed with light drinkers increasing more than moderate/heavier drinkers. However, in the case of females the increase in mastery was just as noticeable among C light drinkers. Even so, E female moderate/heavier drinkers increased more than their C counterparts among whom a very slight and insignificant decrease was seen in mastery.

Fate of Hypothesis 2

The first part of the hypothesis was not supported in that alcohol consumption was found not to co-vary in a linear fashion with distress or with mastery at baseline. In regard to painkiller use, the expected relationship was found with distress but not with mastery. Among users of calming drugs the hypothesis was wholly supported. Among sleeping drug users the hypothesis was supported only to the degree that users reported much higher rates of negatively evaluated life events but little corresponding strain or distress.

The second part of the hypothesis was supported to the extent that moderate users of alcohol, particularly among males, were found to manifest the greatest improvement on most measures of strain and distress. It was among this group of drinkers that the greatest decrease in drinking was seen. However, this should not obviate the importance of gains made by heavier drinkers in stress reduction even though their consumption did not decrease.

The hypothesis was supported in relation to painkiller use in much the same way. There was a strong relationship between decreased use of pain-

Table 6–15
Mastery and Alcohol Use (Q–F Index)

	Mastery Pre-test	Mastery Post-test	Δ	
Experimental Males				
Q–F 1	21.3	22.2	.9	T = −1.30 P > .22
Q–F 2	23.2	24.6	1.4	T = −1.17 P > .27
Q–F 3	22.3	24.2	1.9	T = −2.38 P > .04
Experimental Females				
Q–F 1	20.4	21.9	1.5	T = −1.88 P > .08
Q–F 2/3	23.3	24.3	1.0	T = −0.98 P > .35
Comparison Males				
Q–F 1	23.4	24.0	.6	T = −0.74 P > .50
Q–F 2	22.4	23.2	.8	T = −1.63 P > .18
Q–F 3	23.3	23.6	.3	T = −0.23 P > .79
Comparison Females				
Q–F 1	21.9	23.5	1.6	T = −2.57 P > .02
Q–F 2/3	24.6	24.3	−.3	T = 0.28 P > .81

killers and decreased distress, but this appears to have been largely an artefact of moderate users cutting back their use rather than of changes among heavier users.

The third part of the hypothesis was supported in the general context of the foregoing remarks since, again, the greatest improvement in tension reduction drinking was seen among moderate alcohol consumers both male and female. There appears to have been no relationship between changes in tension reduction drinking and mastery.

Changes in Satisfaction and in Ability to Relax. At baseline, Es tended to be less happy with the kind of work they did and their eating and sleeping habits than Cs. By the time of the second administration, five weeks later, these differences had virtually disappeared. ANOVA significance levels for differences between Es and Cs slipped as follows: pre-test, kind of work, P > .07; eating, P > .06; sleeping; P > .002. Posttest, kind of work, P > .58; eating P > .86; sleeping, P > .57.

It is important to note that changes in self-reported satisfaction with sleep habits are substantiated by reports of symptom reduction on the Derogatis SCL-90-R. Male Es but not females registered significant gains in relation to "trouble falling asleep" (T = 2.41, p > .02), while females but not males registered significant gains in "restless sleep" (T = 3.25, p > .003). These changes were not evident among Cs.

Improvements in the ability to relax (68 percent reported moderate improvement; 32 percent reported major improvement) were in some cases related to above average rates of improvement in other symptom areas. For example, 87 percent of those who reported that their relaxation skills had improved also showed improvement on the SCL-90-R measure of anxiety, as opposed to 60 percent of those who did not feel their skills had improved. Corresponding figures for self-reported improvers versus nonimprovers on SCL-90-R measures were, respectively: hostility, 70 versus 55 percent; trouble falling asleep, 46 versus 25 percent; headaches (from the somatization subscale), 46 versus 25 percent; Positive Symptom Total, 86 versus 70 percent.

Social Support and Strain. At the second administration of the questionnaire, after the course, respondents were asked several questions about the social support available to them (items taken from Brown, Bhrolchain, and Harris [1975]). They concerned the extensiveness and usefulness of the individual's social network as well as the circumstances under which it was used and whether certain associates, friends, or others tended to be selected for discussion of specific problems or simply to share thoughts and feelings. In so far as a specific hypothesis was advanced, we expected that the nature and quality of support would be related to levels of strain and distress. Within the samples available to us in the present study, we found that the principal relationships were between the *extensiveness* of social networks and various measures of distress and strain.

It is interesting that the strongest relationships in the expected direction were found among the C group rather than among the Es, as table 6–16 shows. These relationships are mostly negative, with more extensive support co-varying with lower levels of distress. Not only are there significant correlations between the summative scores of the SCL-90-R (GSI and positive symptom total) and extent of support but also between all the subscales of the same instrument and support. This finding is in keeping with previous research on the value of social support in the maintenance of mental health. What is more surprising, initially, is the absence of similar relationships in the E group. The only exception to this state of affairs is the substantial positive correlation between support and the positive life events score. This suggests either that Es with more extensive support tend to interpret events occurring in their lives as positive or that the fact of extensive support carries with it a preponderance of positively evaluated events. Conversely, people who tend to perceive life in a positive fashion may simply attract a wider group of associates and friends. However, this correlation does not occur among the Cs. This is possibly due to a difference between Es and Cs in relation to the distribution of variance in the extensiveness of support. Among Cs nearly 45 percent claimed to have only one person or no one to call upon in times of stress in contrast to less than 25 percent among Es.

Table 6–16
**Extensiveness of Social Support and Measures of Strain and Distress:
Pearson Correlations**

"Extensiveness" and	E		C	
	r	*Sig.*	*r*	*Sig.*
State Anxiety (Spielberger)			−0.39	.01
Trait Anxiety (Spielberger)			−0.25	.07
Positive Life Events	.40	.001		
Negative Life Events			−0.41	.005
Somatization (SCL-90-R)			−0.28	.04
Obsessive-Compulsiveness (SCL-90-R)			−0.31	.03
Interpersonal Sensitivity (SCL-90-R)			−0.33	.02
Depression (SCL-90-R)			−0.31	.03
Anxiety (SCL-90-R)	.17	.09	−0.36	.01
Hostility (SCL-90-R)			−0.42	.004
Phobic Anxiety (SCL-90-R)			−0.36	.01
Paranoid Ideation (SCL-90-R)			−0.36	.01
Psychoticism (SCL-90-R)			−0.35	.02
Global Severity Index (SCL-90-R)			−0.40	.006
Positive Symptom Total (SCL-90-R)			−0.37	.01
Mastery (Pearlin)			0.23	.09
Tension Reduction Drinking	.16	.09		
Use of Painkillers			0.37	.01

The absence of relationships among Es between support and distress beyond this one exception is nonetheless puzzling. Indeed, if anything there is
a hint of a positive relationship between these variables in that the anxiety
subscale is weakly correlated with extent of support among Es (table 6–16).
Also, tension reduction drinking and support are weakly related. This suggests that higher levels of support are in some minor way related to higher
anxiety and attempts to relieve it by drinking among those who participated
in the relaxation course. Perhaps, however, we are seeing here a reflection of
the phenomenon referred to earlier in which we saw Es responding in an
anxious manner to even *positively* evaluated events. It is possible that interpersonal support, even though considered in a positive light, is also a source
of anxiety to a subgroup of people who enroll in courses of the kind under
discussion here. Other authors (Belle 1982, for example) have suggested that
social support involves a cost–benefit relationship in which one receives
nourishment from others at a certain price. This price may be the felt duty to
reciprocate with understanding and time or more concretely in terms of ser-

vices, money, or goods. In the present case, the cost may be in terms of nervous or emotional energy that is expended in order to maintain otherwise supportive relationships. The anxiety associated with social relationships (supportive though they may be) might have been particularly acute among Es if in fact their perception of the extensiveness of support available to them was inflated by their recent positive experience as members of the relaxation training group.

The possibility that anxiety is positively associated with support (particularly if it is new-found support) among Es may also help to explain why no relationship is found between mastery and support in this group whereas among Cs a weak but positive relationship does exist (table 6–16).

The correlation between painkiller use and support among Cs is positive. No such relationship occurs among Es. Again, this is puzzling. One might expect that pain could be a cost associated with social interaction among Es but not Cs, yet the opposite seems to be true. Currently, we have no explanation for this phenomenon.

Follow-Up

Approximately six months after the course ended, questionnaires were mailed to all participants who had completed first administrations. Twenty-eight Es and fifteen Cs provided usable responses. Unfortunately, ten of the twenty-eight did not supply the requested identifying information that would have permitted linkage between first and third administration data. Consequently, for purposes of this analysis, we have chosen to present quantitative data for only the sixteen Es and fifteen Cs whose data from all three administrations can be matched. For purposes of qualitative data derived from open-ended evaluative questions, however, we have chosen to present the observations of all twenty-eight Es.

Those Es whose third administration questionnaires could be matched to first and second administration data tended to be somewhat older than the typical first administration respondent and were more likely to be female, have children, and be part of families that made more money.

Among Cs no differences of any note were found between third and first administration respondents in terms of demographic and socioeconomic factors.

Results: Quantitative. Clearly any conclusions drawn from a follow-up with a 25 percent return rate must be couched in extremely cautious terms. Superficially, we see improvements that occurred between first and second questionnaire administrations magnified by the time of the third administration. Thus, significant changes became even more significant (table 6–17). Otherwise stated, experimentals became similar to or "better than" comparisons

Table 6–17
Key Scores at Three Administrations

| | Mean Scores | | | | | |
| | Time 1 | | Time 2 | | Time 3 | |
	E	C	E	C	E	C
State Anxiety[a]	42	38	34	38	35	38
Trait Anxiety[a]	42	36	35	36	35	36
G.S.I. (SCL-90-R)[b]	.73	.36	.46	.40	.39	.40
Mastery[c]	22.1	22.1	23.7	22.6	23.5	22.6
Tension Reduction Drinking[d]	12.7[e]	13.8	13.6	14.6	13.6	14.9
n =	16	15	16	15	16	15

[a]Spielberger, Gorsuch, and Lushene (1970)
[b]Derogatis (1977)
[c]Pearlin and Schooler (1978)
[d]Parker and Brody (1982)
[e]Higher scores denote less tension reduction drinking.

by the time of the third administration. However, these observations are subject to the major reservation that we do not know whether the respondents who answered the follow-up questionnaires were representative of those who took the course or even of those who completed the first and second administrations of the research instruments. All that can be said with confidence is that a subgroup of participants demonstrated highly significant improvements. Whether these improvements were attributable to the course is a question that can be partially addressed by reference to the qualitative data presented below. In addition, the question of attribution is considered in the following section through an examination of correlations among changes that occurred among participants between the beginning of the course and the end (five weeks later). We conclude that the pattern of change (correlations among changes) makes it unlikely that improvements were "spontaneous" in the sense of having no identifiable stimulus. We advance the view that the reasonable and probable stimulus was the course itself. Our secondary analysis (high and low normals) confirms this.

Results: Qualitative
Self-Reported Changes. Third administration respondents were asked, "Have you noticed any significant changes in your life during the last six months which you attribute to the course on stress management/relaxation?" Of the twenty-eight replies, all but four were positive. Nearly 60 percent were cer-

tain that they had experienced changes due to the course, while 26 percent said they thought the changes they had experienced were due to the course. The specific types of changes described may be rank ordered as follows:

1. Positive attitudinal and mood changes (37percent)
2. Improved coping ability (30 percent)
3. (Tied) Feeling more relaxed in general (26 percent)
   Health/health habit improvements (26 percent)
4. More energy (19 percent)
5. Ability to recognize stress and stressors (15 percent)
6. Improved sleep (11 percent)

Note that all the responses just listed were generated in reply to an open-ended question asking for a description of the changes rather than in response to a check list. Many respondents cited two (39 percent), three (17 percent), or more (9 percent) of these changes.

Relaxation "At Will". Respondents were asked to rate the course, on a scale of 1 to 10, according to the degree to which it had attained the objective of teaching them how to relax at will—that is, get into a state of relaxation more or less when they wanted to, and quickly. The mean rating was 7.4 among the group as a whole. Among those who attributed personal change to the course with certainty ($n = 16$), the mean rating was 8.2. Among those who were less certain, it was 6.4, and among those who did not attribute changes to the course at all, it was 6.0.

Correspondingly, when asked about the difficulty involved in relaxation, respondents answered that relaxation is now: very easy (15 percent); moderately easy (63 percent); moderately difficult (15 percent); very difficult (7 percent).

Not surprisingly, 62 percent of those who claimed that relaxation was now moderately or very easy rated the course at 8 or above while only 33 percent of those who found relaxation moderately or very difficult rated it at this level.

Current Use of Skills. Respondents were asked whether they still practiced the skills that they learned during the course. Twenty-one percent replied that they still used all the skills, 64 percent that they used some, 7 percent that they had added new skills since, and 7 percent said they no longer used the skills at all. Many reported that they used the skills only when needed (38 percent), others that they used them up to two or three times a week (31 percent). Thirty-one percent reported daily or almost daily use.

Twelve of the current users (46 percent) said that they had modified the skills to some extent in order to suit their own needs or habits. It is interesting to note that those who reportedly modified the skills were the most frequent users of them.

Relationships among Observed Changes. In the process of presenting the results of the evaluation we have made reference to relationships among the changes that were observed. These are summarized in Figure 6–1. While the presentation of these relationships is not meant to imply causality, it is perhaps reasonable to infer that the catalyst for many of the noted changes was improvement in the ability to relax. Relaxation appears to have been the key that provided access to a variety of other changes, but it is difficult to determine in what order they took place. For example, it is unclear whether the ability to relax had a direct impact upon trait and state anxiety and upon distress (the GSI and PST superscores of the Derogatis SCL-90-R) or whether it was mediated through improvements in mastery (self-efficacy). Similarly, it is unclear whether the use of alcohol, painkillers, and coffee changed as a direct result of improvements in the ability to relax and reductions in strain and distress or whether these changes were mediated by improvements in the amount and quality of sleep. A similar question can be raised about improvements in satisfaction with work habits and eating practices.

There is a sense, of course, in which all these changes are interdependent. For example, if by sheer force of will one could modify one's consumption of alcohol, coffee, and painkillers and could consume less food, less often, and under more restful circumstances, it is possible that one would sleep better and feel better. Further, one might feel more in control of life (or at least of personal health) and as a result feel less anxious and distressed. In short it might be possible to begin at the behavioral end of the problem and work back to improved mental health.

Such a reverse process, however, though not uncommon when the problem to be corrected is not too severe, is difficult to implement when things have piled up on a person to the extent that he or she feels trapped by the enormity of changes that need to be made. Relaxation training appears to provide a leaven of well-being that activates the individual's ability and will to break out of the vicious cycle of deleterious habits that otherwise tend to reinforce each other endlessly. One of the chief characteristics of distress is that people subject to it tend to lose perspective on their problems. After a certain point they tend to compound their difficulties with half-thought-out solutions that become part of the problem. Relaxation training can provide for a significant group of people a means of centering themselves in the benign sense of allowing them to marshall their personal resources and consider from whence they came and whither to proceed. It permits the possibility of at least some measure of self-reintegration at a happier and healthier level. In

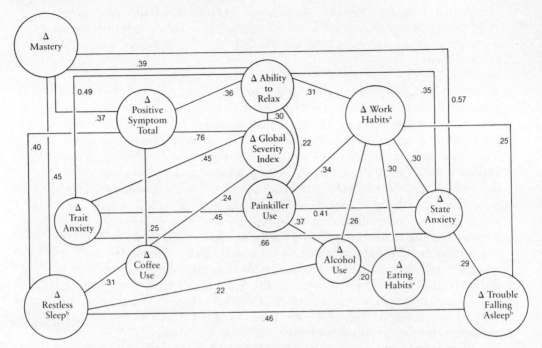

Note: This is not a path diagram, simply a graphic representation of correlations as currently understood. Read changes as "improvements."

[a]Full titles: "Satisfaction with Work Habits, Eating Habits."

[b]Not shown: correlation between Δ Restless Sleep and Δ Global Severity Index = 0.36; Δ Trouble Falling Asleep and Δ Global Severity Index = 0.21.

Figure 6–1. Principal Correlations among Changes Observed in the "Beyond Stress" Groups (*n* = 63). Signs omitted.
(all correlations are at or above the .05 level of significance)
Δ = change, pre- to post-course.

so far as relaxation appears to have systemic life-style effects, it may be thought of as a vital component in a holistic approach to mental and physical health.

Case Study 2. Evaluation of "Creative Coping with Stress": A Stress Management Course for Teachers

Teaching has come to be regarded as a highly stressed occupation (see Dunham 1976; and Friedman, Lehrer, and Stevens 1983 for a review). Many deficits have been attributed to high stress among teachers. Among them are less effectiveness in the classroom, increased student anxiety, lower student achievement, poor teacher–pupil rapport, and inability to maintain discipline (Friedman, Lehrer, and Stevens 1983).

To a great extent it seems to be assumed that the source of this stress can be found in the classroom and general working conditions associated with teaching in modern times, and not with the teachers themselves (Sutton 1984).

In an earlier survey conducted by the authors about stress, strain, and coping carried out among the membership of District 10, Ontario Secondary School Teachers' Federation (OSSTF), we found that there tended to be a significant relationship in many cases between stress experienced at work (in school) and stress and dissatisfaction experienced in other primarily domestic and social situations (Shain, Boutilier, and Simon 1984). Here, stress was defined in relation to positively and negatively evaluated "life events" to which respondents had been subject in the last year or six months. Such events were both general in nature, based on the earlier work of Holmes and Rahe (1967), and also specific to teaching. For the latter items we drew upon work conducted by the OSSTF and provided to us as a list of teaching-specific events (Bergey 1983). These items overlap to a considerable degree with those used by Sutton (1984).

It was also found that as stress increased subjectively experienced strain (as measured by Spielberger's State and Trait Anxiety Scale) also tended to increase. However, it was observed that in some cases people who experienced high stress were not unusually strained while some who reported low stress were very highly strained. Also, a very strong relationship was found between high strain and low sense of mastery (Pearlin and Schooler 1978).

In addition to predicting classroom deficits of the kind outlined above, high strain has been shown repeatedly to be associated with greater susceptibility to a wide range of morbid physical and mental health conditions (Selye 1976; Dohrenwend and Dohrenwend 1974).

Efforts to reduce strain among teachers have been reported as successful in some instances (Friedman, Lehrer, and Stevens 1983), just as efforts to do so among other normal, working populations have been shown to be successful (Schwartz 1982). In the case reported here, a contract was established between representatives of the OSSTF (District 10) and two local agencies, the Peel Family Service Association and the Addiction Research Foundation of Ontario for purposes of developing a stress management course that was offered under federation auspices.

The evaluation was conducted in order to determine (a) the impact of the particular blend of components used in this course on reduction of strain, (b) the relationship between reduced strain and reduced consumption of alcohol, painkillers, coffee, and tobacco use, and (c) the impact of reduced strain on mastery.

Hypotheses

1. A stress management course for teachers will be effective in reducing strain among volunteer participants to levels similar to those found in a comparison group who did not take the course.
2. Among those who drink at above moderate levels and who experience high strain there will be significant reduction in alcohol and drug use as a function of reduced strain.
3. Reduced strain will lead to increased mastery.

These hypotheses are based upon a conceptual model of stress and strain in which the impact of external events (stressors) is seen to be mediated or "buffered" by a number of factors, among them being a sense of mastery (Roskies and Lazarus 1980; Bandura 1977a; Pearlin and Schooler 1978) and possibly alcohol and drug use, including tobacco (Marlatt, Kosturn, and Lang 1975; Marlatt 1976). In this model, alcohol and drug use are seen as reinforcing the individual's sense of mastery through a variety of physiological, psychological, and social mechanisms, and perhaps related to this, alcohol and drug use may act to reduce tension physiologically or may be engaged in with the *expectation* that it will reduce tension as noted in Part 1 of this chapter. These alternative conceptions of the role of alcohol and drugs may not in fact be in competition, a possibility which gathers some support from the results of the present study.

The Intervention

"Creative Coping with Stress" comprised four sessions of approximately three hours each. There were three main elements of the course. The first was *cognitive input* via "mini-lectures" on: definitions of stress and the stress process; sources of stress; range of coping responses (functional/dysfunctional); outline of skills that may be learned in order to cope more effectively with stress; factors that mediate stress; stress response-style as learned behavior; influence of beliefs on perceptions of stress; transactional analysis as a framework for comprehension of interpersonal stress; and time management as a stress-reducer.

The second was *skill development,* involving exercises related to: identification of sources of personal stress; identification of personal coping responses; relaxation; and meditation.

Group interaction/support was the third element, involving group exercises in stress-source identification, relaxation, self-disclosure, and sharing of experiences. Participants were asked to contract with other members of the group and with the group leaders to participate fully, to attend regularly, and to work at the assigned tasks, including homework.

Outline of Skills

Personal Skills	Interpersonal Skills
Deep muscle relaxation (DMR)	Relationships with partners,
Progressive muscle relaxation (PMR)	spouses, children, co-workers, supervisors, subordinates
Meditation	Definition of problem ownership
Relaxation response	Problem solving
Imaging	Constructive confrontation
Yoga	"I" messages
Specific breathing exercises	Assertion
Giving self "strokes"	Acceptance of immutables
Exercise	Providing "strokes" to others
Recreation (including hobbies, distractions, fantasies)	Active listening

Most of these skills were practised to some extent during the course. However, it appears that in most cases they were illustrated through one or two practice sessions rather than developed to a high level of proficiency.

Course Participants. Volunteers were recruited from the membership of the OSSTF in District 10. Advertisements were the primary means of solicitation. Twenty-two people initially signed up for the course; seventeen of these completed a voluntary pre- and post-course evaluation questionnaire and eight of these completed a third questionnaire six months later. Of those who completed the first two questionnaires 24 percent were men and 76 percent women; 31 percent were single, 56 percent married, and 13 percent had common law arrangements. Thirty-five percent had children. The mean age of participants was just under 37 years (range 28–50). Ninety-four percent had a university education; the others had community college or partial university training. Fifty-three percent said their combined family income was $55,000 a year or more. Ninety-four percent were born in Canada. As regards previous exposure to stress-related activities, 35 percent had taken courses, 76 percent had read books or articles, and 59 percent had seen films or heard radio programs. They cited the reasons for taking the course as: job stress (35 percent), health (29 percent), general, personal stress (100 percent), and desire to improve performance (6 percent).

Comparison Group. The same Cs were used in the evaluation of "Creative Coping with Stress" as were involved in the evaluation of "Beyond Stress" described in the first part of this chapter.

Once again, the design of the study was quasi-experimental in the sense that the Cs and Es were not randomly allocated from an eligible pool into a control group that received no training and an E group that did. The Cs in this study, as in the first, were considered mainly as a reference group of normals.

The principal test of the first hypothesis was carried out according to the expectation that, if the course had been successful, Es (teachers) would become more like the Cs on key mental health variables by the end of the course.

The C group closely resembled the teacher Es in age (mean, 38 years). However, they were less educated, with only 37 percent having university education, although 75 percent had completed high school. Their family income was not as high as the Es, with only 32 percent earning more than $55,000. About the same proportion of Cs as Es were married, but a larger proportion of the former were divorced (8 percent), separated (10.5 percent), remarried (5 percent), or widowed (2.6 percent). It is interesting to note, however, that 18 percent of the Cs had previously taken stress courses.

Measures and Results

The questionnaires used in this study were identical to those in the evaluation of "Beyond Stress." Findings pertaining to the first and second administrations are presented separately from those pertaining to the third since the sample base for the latter is smaller than for the former.

Changes from Pre- to Post-course (First to Second Administration)
Relationship of Positive to Negative Stressors. The pre- and post-comparisons shown in table 6–18 are based on different time referents: the pre-course evaluation score is based on the twelve months prior to the course, while the post-course evaluation score is based on the four weeks of the course itself. Even so, there may have been at post-course some reevaluation of events that took place in the year prior to the course. This phenomenon was also noted in case study 1.

The most noticeable feature of the figures shown in table 6–18 is the difference between Es and Cs at pre-course with regard to the ratio of negatively evaluated to positively evaluated events. Among Es, negative events outnumber positive by more than 2:1, while among Cs the ratio is closer to equivalence. At post-course, this situation no longer prevails. Positive and negative events are reported with almost equal regularity among both E and C groups. Readers may recall that similar observations about the relationship of positive to negative stressors were made concerning the participants in "Beyond Stress." These data are open to more than one interpretation. On the one hand, Es may simply have had an unusually bad year prior to the course, and this fact may have been a catalyst to their registration in the course. On the other hand, Es may be predisposed to evaluate events in a

Table 6–18
Life Experiences Survey:[a] Changes in Scores Pre- to Post-course (Teachers)

Evaluation Scores	Events in Previous Year: Pre-course		Events in Last Four Weeks: Post-course	
	E	C	E	C
Positive Total	5.1	5.4	3.1	2.7
Negative Total	11.2	6.1	3.3	2.7
Grand Total	16.3	11.5	6.4	5.4

[a]Sarason, Johnson, and Siegel (1978).

negative light, or they may behave in such a way as to engender negative events. The evidence we have suggests support for the former interpretation. During the time of the course, one would have expected the same disproportionality of negative to positive evaluations to have prevailed if the participants were disposed toward negativity, which did not occur.

Spielberger State and Trait Anxiety. Surprisingly, the change in trait anxiety among Es observed in table 6–19 was more statistically significant than the change in state anxiety. However, a trend toward improvement in state anxiety is discernible: the change in mean score from first to second administration is the equivalent of a shift from scores that characterize Spielberger's normative sample of neuropsychiatric patients at the 66th percentile to scores for the same sample at the 41st percentile. Even so, this change still leaves the participants at a level characteristic of university undergraduates at the 86th percentile.

With regard to trait anxiety, the change in mean scores from first to second administrations is equivalent to a shift from scores that characterize neuropsychiatric patients at the 45th percentile to those at the 33rd percentile (70th percentile for university undergraduates). The difference in significance levels may be accounted for by reference to the difference in standard deviations: in the case of state anxiety the spread of scores is just as large after the course as before (although about a lower mean), while in the case of trait anxiety the spread of scores diminishes as well as the mean. Nevertheless, it is unusual to see change in a relatively stable characteristic such as trait anxiety outstrip changes in a situation-specific response pattern such as state anxiety.

Thus, between the first two administrations we see a normalizing trend relative to the C (reference) group, but a trend that nonetheless leaves the Es still more anxious than their nonparticipant counterparts. However, another type of comparison may be in order. In our survey of teacher stress, strain,

Table 6–19
State and Trait Anxiety:[a] Changes in Scores Pre- to Post-course (Teachers)

| | E | | C | |
| | Combined Males/Females | | Males | Females |
Administration #	(n = 16)		(n = 17)	(n = 21)
State Anxiety Raw Scores[b]				
1	53		33	42
	(S.D. = 15.65)			
2	46		35	40
	(S.D. = 15.50)			
Trait Anxiety Raw Scores[c]				
1	45		35	37
	(S.D. = 10.73)			
2	42		36	37
	(S.D. = 9.73)			

[a]Spielberger, Gorsuch, and Lushene (1970).
[b]*T test* (Es only) T = 1.62, P > .13
[c]*T test* (Es only) T = 2.39, P > .03

and coping referred to earlier (Shain, Boutilier, and Simon 1984), we found the mean score of teachers on state anxiety to be 40 for males and 44 for females while for trait anxiety the mean scores were 34 and 36 respectively. These scores are higher than the Cs' in the case of state anxiety but about the same in the case of trait anxiety. In the Friedman, Lehrer, and Stevens study (1983), the baseline state anxiety score was 44 (S.D. 12.02) and five weeks later it was 34 (S.D. 9.81). The corresponding figures for trait anxiety were 41 (S.D. 10.54) and 34 (S.D. 8.30). Thus, while the changes were somewhat greater than in the present case, the baseline levels were not as high in the Friedman, Lehrer, and Stevens study.

Derogatis Symptom Check List (SCL-90-R). For present purposes we shall refer only to the summative, grand scores generated from the SCL-90-R. The two scores in question are the Global Severity Index (GSI), already described, and the Positive Symptom Total (PST). The PST is simply a count for each individual of all those items that were given a non-zero score. Both summative scores express in somewhat different ways the extent to which individuals are generally "distressed" by a variety of symptoms the putative origin of which is psychological.

The change on the GSI from first to second administration observed in table 6–20 is roughly equivalent to a shift from scores characteristic of Der-

Table 6–20
Derogatis Symptom Check List:[a] Changes in "Superscores" Pre- to Post-course (Teachers)

	E		C	
Administration #	Males/Females Combined (n = 16)		Males (n = 17)	Females (n = 21)
GSI Mean Raw Scores[b]				
1	.80		.24	.44
2	.62		.28	.48
PST Mean Raw Scores (Sexes Combined)[c]				
	E (n = 16)			C (n = 38)
1	40			25
2	34			23

[a]Derogatis (1977).
[b]T test T = 3.96, P > .001 (Es only)
[c]T test T = 3.06, P > .01 (Es only)

ogatis's normative "normal" population (in this case female) at the 90th percentile to scores characteristic of the same group at the 84th percentile. While this may not appear to be a large change, it does in fact bring the sample mean down to the borderline between subclinical and normal score ranges whereas at pre-test it was squarely in the subclinical range (1st standard deviation). Again, as in the case of the Spielberger Anxiety measures, this represents a normalizing trend relative to our C group whose scores *increased* slightly but not significantly during the same period.

Use of Alcohol, Painkillers, Coffee, Tobacco. A quantity–frequency (Q–F) index was constructed from the data as in case study 1 so that a single score reflected the amount typically consumed and how often. In this index, drinking occasions were coded as follows: none; once per month = 1; 2–3 times per month = 2; once per week = 3; 2–3 times per week = 4; more than 3 times a week but less than daily = 5; daily = 6. This code was multiplied by the actual figure recorded by respondents as representing the amount consumed on a typical drinking occasion. The product of this multiplication could convey a variety of meanings, for example, "12" could be derived from 6 × 2, which means daily drinking at the rate of 2 drinks per occasion; or it could be derived from 4 × 3, which means drinking 2–3 times a week at a rate of 3 drinks per sitting. However, we found that the index served quite well as a single measure of consumption, correlating well with individual measures of frequency and amounts.

Baseline Q–F scores for the two groups were quite similar (E, 6.31; C, 6.90). Little change was seen after the course (E, 6.25; C, 6.80). The baseline mean scores were low, reflecting an average consumption pattern equivalent to just over one drink per day.

On closer inspection, it appears that a decrease in consumption among 31.4 percent of Es was offset by an increase in consumption among another 12.5 percent (two people). Even though a similar trend occurred among Cs, a phenomenon occurred among Es that was unique, namely a correlation between decreased consumption of alcohol and increased mastery scores (R = -0.53, p > .02). Thus, an apparently random change in alcohol use assumes greater significance as a result of its being reinforced by a change in perceived self-efficacy among teachers who participated in the course.

No changes of any practical or statistical significance were seen in the areas of painkiller, coffee, or tobacco use.

Change in Reported Ability to Relax. Prior to the course, 47 percent of the teachers who took part in it reported that relaxation was, for them, very difficult or impossible. Subsequent to training, only 18 percent still reported this level of difficulty. Among those claiming that relaxation was easy or moderately easy, the proportions were reversed, with 18 percent so reporting prior to the course and 47 percent afterwards. While this reflects a normalizing trend relative to the Cs, the teachers as a whole still remained less able to relax than their nonparticipant counterparts, 73 percent of whom said, on follow-up, that they found it easy or moderately easy to relax.

Mastery. Teachers' mean scores on mastery increased significantly during the course (table 6–21). The changes were "patterned" in that they correlated to a significant degree with other important changes (note: table 6–21).

These relationships are important in that we can see a change in self-efficacy, possibly resulting from reductions in distress, anxiety, and depression. However, the course had a number of elements that had a direct cognitive appeal to sense of self-efficacy, so that it is arguable that some reduction in distress resulted from improvements in sense of control. Probably the relationship is of a synergistic or reciprocal nature, with improvements in one area catalyzing changes in another, which in turn reinforce and augment changes in the former.

It will be recalled that a correlation between change in mastery and decrease in alcohol consumption has been already noted (R = 0.53, P < .02).

Follow-Up (Third Administration). Eight of the seventeen participants who completed the first two administrations also completed the third carried out

Table 6–21
Mastery[a] Scores: Pre- and Post-course (Teachers)

| | Mean Mastery Scores | |
Administration #	E (n = 17)	C (n = 38)
1	21.0	22.1
2	23.4	22.6

T = 3.49, P > .003 (Es only)

Note: *Correlates of Changes in Mastery[a] (Teachers)*
Δ = Change from pre- to post-course.
Δ Mastery[a] and Δ Global Severity Index[b] R = −0.53 (P > .02)
Δ Mastery and Δ Positive Symptom Total[b] R = −0.61 (P > .01)
Δ Mastery and Δ Depression[b] R = −0.58 (P > .01)
Δ Mastery and Δ Trait Anxiety[c] R = −0.58 (P > .01)
[a]Pearlin and Schooler (1978).
[b]Derogatis SCL-90-R (1977) GSI, PST and Subscale
[c]Spielberger, Gorsuch, and Lushene (1970).

six months later. All were female and almost all were married, though only a quarter had children. Beyond this, the characteristics of these women were identical to those who did not complete the follow-up.

Quantitative Results. The women in the follow-up sample appear to have consolidated gains that were made during their course on stress management. It is important to note that the baseline (first administration) scores of this group were very close to those of females as a whole at that time, that is, they were neither more nor less distressed than their colleagues in the course. Even so, there is no safe basis upon which to generalize from the eight women at follow-up to the thirteen women who completed only first and second administrations. Nonetheless, it is clear that at least a sizable subgroup (62 percent) of the thirteen women continued to improve during the six months following the course (table 6–22). In the cases of trait anxiety and distress as measured by the GSI of the SCL-90-R, the follow-up women closely resembled the Cs, while in relation to mastery they appeared even better than Cs at follow-up.

Qualitative Results. Former participants in "Creative Coping with Stress" were somewhat hesitant to ascribe to the course changes that they had undergone since their training. Sixty-three percent said that "possibly" the changes that they had experienced during the preceding six months were attributable to the course, while only 25 percent said "certainly." However, respondents were for the most part able to say in answer to an open-ended question that

Table 6–22
Key Scores at Three Administrations (Teachers)

	Mean Scores					
	Time 1		Time 2		Time 3	
	E	C	E	C	E	C
State Anxiety[a]	53	38	45	38	44	38
Trait Anxiety[a]	45	36	41	36	37	35
GSI (SCL-90-R)[b]	.84	.36	.57	.40	.36	.40
Mastery[c]	21.8	22.1	24.1	22.6	25.1	22.6
Tension Reduction Drinking[d]	12.1[e]	13.8	12.8	14.6	12.9	14.9
n =	8	15	8	15	8	15

[a]Spielberger, Gorsuch, and Lushene (1970)
[b]Derogatis (1977)
[c]Pearlin and Schooler (1978)
[d]Parker and Brody (1982)
[e]Higher scores denote less tension reduction drinking.

positive attitudinal and mood changes were among those changes most likely to have been due to the course (86 percent). Forty-three percent referred to "improved coping" as an identifiable benefit. Five of the eight women claimed that they continued to use the skills taught during the course, four of these "when needed" rather than routinely. The same four had added their own touches to the methods they had learned, a process recommended by the course organizers. The ability to relax at will was not a specific objective of "Creative Coping with Stress" in the same way that it was in "Beyond Stress"; nonetheless, the eight follow-up participants ranked the course on the average as six out of ten on this dimension.

Fate of the Hypotheses. 1. This hypothesis was partially supported in that strain among participants, according to the Spielberger and Derogatis instruments, was significantly reduced but not to the level exhibited by the C group. Participants also reported that they found relaxation considerably easier after the course.

2. The second hypothesis was not testable with the present group of participants because they were small in number and contained few heavy drinkers. Nonetheless, some evidence of reinforcement of moderate drinking practices was produced, although it should be considered as no more than suggestive. Even so, the potential for stress management programs in reinforcing moderate drinking practices seems worthy of further investigation.

3. The third hypothesis was supported. There were significant gains in mastery among participants and these gains were correlated with a decrease in strain.

Follow-up data suggest that, at least among a subgroup of participants, the noted changes gather rather than lose strength over time. We cannot with any confidence generalize from this group to all the participants. However, we should note that third administration completers were as much distressed at baseline as their fellow participants. This fact at least argues against the possible objection that follow-up participants were less in need of help to begin with.

If we consider the observed changes reported here in their totality we have a whole that is probably greater than the sum of its parts. Relatively modest, but statistically significant reductions in distress are seen to be related to similarly modest improvements in mastery: taken together, however, their clinical significance may exceed that of either change considered in isolation.

The relationship between reduction in distress and improvement in mastery is a very encouraging phenomenon. It suggests that a course of the type described here can help to interrupt or reverse a negative chain of events in which self-efficacy is eroded by stress and strain and where further despondency about the ability to influence events leads to yet more distress. This precarious balance between strain and self-efficacy has been referred to frequently in the literature but has much less often been shown to be amenable to change.

A Comparison between "Beyond Stress" and "Creative Coping with Stress"

In this chapter, two courses in stress management were described and the results of their evaluations reported. "Beyond Stress" focused a great deal upon the teaching of relaxation skills, while "Creative Coping" emphasized cognitive skills.

It is perhaps not surprising, since "Beyond Stress" concentrated so much upon skill training and included homework assignments related to it, that the impact of this program was greater in relation to key measures of distress. However, direct comparisons of this kind may be inadvisable, since the participants were more homogenous in "Creative Coping" and also tended to score higher at baseline on key measures of distress than those in "Beyond Stress." The sample base in the latter course was also considerably larger than in the former, permitting analyses that would have been impractical in smaller numbers. For example, the impact of training upon alcohol and drug use could not be investigated according to risk groups based on Q–F scores among "Creative Coping" participants for this reason.

A striking convergence of results can be seen in the relationships among the changes that were attributed to both courses. The connection between self-efficacy and distress in terms of reciprocal influence is particularly noteworthy, since it has implications for future design of stress programs. Direct appeals to self-efficacy may be of greater importance than they are often assumed to be. It appears possible that important gains in mental health may be engendered from gains in this realm, which in turn may result from direct cognitive appeals. Conversely, gains in self-efficacy appear to result from training in specific relaxation skills.

It seems premature and unwarranted to make claims for the superiority of one approach over another based on the evidence reported here. However, the relative efficacy of various stress management techniques all need to be seen from the perspective of potential participants who probably would wish for a *choice* of approaches so that they could select one or more that suited their needs and preferences.

7

The Synergism of Employee Assistance and Health Promotion Programs: Cost–Benefit Considerations

I n this chapter we present a case for the view that the cost–benefit ratio for employee assistance programs (EAPs) and health promotion programs (HPPs) as separate program entities is tipped in favor of "benefit" when the two are planned and implemented as parts of a comprehensive approach to the mental and physical health of people in the workplace.

The term cost–benefit has perhaps obscured more than clarified the relationships that may exist between the resources that are expended to produce results and the results themselves. This may be because there are at least four major conceptual models of cost–benefit as applied to our area, probably more. However, the ones that we distinguish here each provide a somewhat different answer to the question, "Why are we concerned about improving the health of workers through either EAPs, HPPs, or indeed any other kind of intervention?"

Four Cost–Benefit Models

The Cost Recovery Model

At its simplest, this version of cost–benefit requires that the cost input to a program be recovered (a breakeven position) or that the "saved" or "recovered" costs exceed the program costs. For example, if it is known that installation and maintenance of an EAP for two years will cost $25,000, then it is anticipated that the value of reduced absenteeism, medical costs, workers compensation disbursements, and so on will exceed that amount when computed as the difference between costs represented by referred employees before and after treatment. We should recognize, however, that a minority of employers keep records with sufficient accuracy to permit this type of accounting.

The Purchase of Service Model

While there is a concern here about receiving value for money, the concern is more likely to be manifested by requiring the service provider (often a third party outside the organization) to account for how his or her costs are computed and by requiring that these costs be reasonable in relation to other means of providing the same service. Thus, it is a relative rather than absolute form of costing in a competitive market environment. Benefits are either assumed in this model or are deduced from softer evidence than would be the case in the cost recovery model; for example, attendance and participation rates in EAPs are frequently taken as evidence that benefits are accruing. The service being purchased may have perceived value in a number of unpriceable areas, among which are the good will of the work force, the image and reputation of the corporation, confidence of investors both present and future, and recruitment of new staff.

The Benefits Model

According to this model, the costs of program start-up and maintenance (either EAP or HPP) are considered to be part of the benefits or compensation package offered to employees. In this sense, program costs may be written off as costs for which general rather than specific returns are expected. Undoubtedly, program benefits serve to attract better people and keep them longer, but no one expects to establish a clear causal path between program implementation and honing a competitive edge. There is anecdotal evidence to suggest that some industrial insurance companies reward organizations with EAPs and HPPs by reducing health coverage premiums. This is currently a gray area since some types of program may actually *increase* the utilization of insurance. While this may be fully anticipated in the short run, the concern is related more to chronic use of health care benefits through self-referral. In any case, long-term studies are required to establish whether, over the long run, costs to carriers increase. It may be that the rate of utilization goes up with self-referrals but the cost per unit (visit) diminishes. One might expect such a trend if preventive self-care is taking the place of remedial care. Of course, it has been pointed out too that keeping good people alive and well longer may be a drain on the pension fund (Faust and Vilnius 1983, 526).

The Ecological Model

This concept of cost–benefit emphasizes the conservation of vital resources, in this case human resources. It appears to be predicated on the notion that employers should seek to replenish the stores of mental and physical energy that they draw from their workers by means beyond simply paying them for their time and labor. In a sense, this is an extension of the benefits model, but it goes further in that it is often found in the presence of a philosophy of doing work in which individual health, organizational health, and social health (the fitness of the nation) are inextricably linked. There is as much concern about the reduction of factors in the work environment producing superfluous stress as there is about the provision of stress management courses for those in need. In this philosophy, it is no more appropriate to expose employees to hazardous levels of superfluous stress than it is to expose them to toxic chemicals or dangerous equipment. In this model, then, cost is what you can afford, and benefit is the assumed value of treating workers with respect and dignity. In this context, cost is as amorphous a concept as benefit since it may involve not only financial outlay but also (and perhaps more importantly) a rethinking of how work is organized and carried out. This is not a new concept; indeed, within the socio-cultural and economic context of the times, one is able to locate examples of this ecological outlook throughout history. It has never been normative, but it has rarely been entirely absent.

Application of the Cost–Benefit Models

The published literature on evaluation of EAP and HPP when discussing cost–benefit analysis seems to be preoccupied with the cost recovery model with some little attention to the purchase of service model and hardly any to the benefits and ecological models. However, when talking to executives about their programs, the emphasis tends to be quite different: often it is the benefits model that receives prominence followed by the purchase of service and ecological perspectives (see Roman 1981a; Sapolsky et al. 1981; Groeneveld et al. 1984). There are vivid exceptions to this, of course, but the concern about cost recovery seems to *precede* program initiation and then diminishes once the program is underway. Concern varies, too, according to the amount of investment in the program. Clearly, when special staff are hired and given an operating budget in order to run a program, greater accountability can be expected than in the case where an existing position is extended to include program coordination. Perhaps this concern is justified given the equivocal evidence to date on cost recovery alone.

EAPs and Cost–Benefit

Typically, EAPs have been evaluated from two points of view, whichever cost–benefit model is applied, since the importance of variables within the models are matters of emphasis and degree rather than kind. The principal dimensions are *coverage* and *effectiveness* (see Shain and Groeneveld 1980 for a more thorough review).

Coverage really refers to what is often called the *penetration rate*. At the formal level this means the rate at which problem employees are identified by others and managed according to the established procedures of the policy, if there is one. This frequently means referral, so *referral rates* become another dimension of program performance. At the more informal level, coverage refers to *utilization rates*—the rate at which workers invoke the program on a voluntary or quasi-voluntary basis. This may involve self-referral or simply the acquisition of advice.

At the effectiveness level, we are dealing with a broad spectrum of potential program-outcome variables among which are: job status and job performance; status of the presenting problem and related problems; rate of disciplinary actions; rate of accidents and workers' compensation (WCB) claims; use of sick leave, benefits; use of grievance procedures; and rate of unauthorized absences. The status of the presenting problem and related problems can be broken down to a fine degree, involving a thorough inventory, before and after intervention, of psycho-social functioning at work and at home.

Essential to the notion of effectiveness, of course, is the period over which the program is evaluated. That is, if benign results occur, how long do they last? This determines the *survival rate*.

In both coverage and effectiveness studies we are interested not only in "How many?" but also in "Who?" That is, what are the characteristics of those people who come to the program as opposed to the characteristics of those who do not even though they may be problem employees or employees somehow in trouble?

These coverage and effectiveness dimensions of program performance are related in the literature to the huge range of program types. They vary enormously in quality and in mode of operation, in scope or breadth and in allocated resources. Nevertheless they are often compared. It is no wonder that the results are so varied and apparently contradictory. In fact, there is a real problem in obtaining much beyond coverage data for any problems other than those related to alcohol. Even though broad-based EAPs have been dealing with a wide variety of problems for some years, there seems to have been a disproportionate amount of interest in their ability to manage alcoholics. In so far as coverage data do exist, they suggest that achieved identification rates are considerably below the rate at which the problems in question prevail in the workplace. In the case of alcohol-related problems, for example, Hitchcock and Sanders (1976) estimated from a review of published studies that between 0.6 and 1.6 percent of the work force were likely to be identified. The programs which they were studying tended to be of the alcohol-specific variety. Their estimates took into account, where possible, the total life span of these programs. In a more recent study (Shain, 1985), it was shown that even when a sample of broad-based EAPs is taken as the referent, the referral rates for alcohol-related problems fall into almost exactly the same range, with the same mean and median. In the same study, the coverage rates for all problems combined were also investigated in relation to ten broad-based programs. Referral rates fell within a range of 0.56 to 4.6 percent with the mean at 2.15 percent.

It is instructive to compare these referral rates with estimates of the prevalence of the problems in question. In the studies reported in chapter 3, it was shown that prevalence rates of alcohol-related problems fell within the 9 to 15 percent range (when problem drinking and presumptive problem drinking were combined) while mental health problems, defined according to high scores on a cumulative "distress" index often fell into the 15 to 20 percent range. Although much of the interpretation of such figures depends upon the criteria used to define the problems, even when the strictest criteria are used the prevalence of problems far exceeds the actual referral rates in most EAPs. With regard to alcohol-related problems, our prevalence estimates are not out of keeping with those of Mannello and Seaman (1979) and Parker and Brody (1982). With regard to mental health problems generally, the studies of Weiner, Akabas, and Sommer (1973), the President's Commission on Mental Health (1978) and Masi (1982) support our estimates of their prevalence.

With regard to the second major evaluative dimension, effectiveness, the data do not exist in any coherent form with regard to problems other than alcoholism. Anecdotal accounts abound, but the optimistic note struck in these testimonies has yet to receive harmonic support from well-controlled studies that investigate the outcome of intervention through EAPs over long periods of time. The evidence suggests that success rates for the treatment of alcoholism are high relative to those found in general or psychiatric hospital patient populations. However, there is reason to believe that these rates, often reported to be between 65 and 80 percent (with huge variations in follow-up), are somewhat inflated by the prejudicial use of employer criteria in defining success (Shain and Groeneveld 1980; Walker and Shain 1983). These criteria tend to emphasize the *manageability* of problems among employees rather than their *resolution* in psycho-social terms. Thus, within this employer-oriented framework, success can include resignation, termination, transfer, demotion, "backwatering," medical retirements, and early retirements. We hasten to add that these criteria are legitimate in themselves: our proviso is that such criteria should not be confused with those in the psycho-social domain involving such dimensions as alcohol involvement, interpersonal/familial/domestic functioning, anxiety, depression, and the wide variety of other mental health conditions that can lead to actual or potential job performance problems. Recent research by Groeneveld and associates (1984; 1985) adds fuel to the argument that the addition of robust after-care components into EAPs could substantially alter the success rates of these programs in the psycho-social area. The absence or poor quality of follow-up was one of the main areas of concern indicated by EAP service consumers in a recent study of forty-five such programs in metropolitan Toronto reported in chapter 4 of this book.

Leaving aside the finer questions of evaluation for the moment, it will be apparent from the foregoing that EAPs appear to have marked limitations with regard to the proportion of the population at risk that they can reach. Once having reached what we might assume are the highest risk candidates, the programs appear to do well in relation to alcohol problems if we are prepared to accept a certain bias in the definition of success. With regard to other problems, we have little or no precise information about success rates.

This is not an indictment of EAPs. It is likely, as other writers have suggested, that at any one time it is a very small proportion of the population that is at risk in psycho-social terms, which accounts for a disproportionately large amount of the loss associated with alcoholism and other mental or physical health problems. This small group has earned the title "critical few" in the literature on loss control management (see, for example, Bird and Loftus 1976). Borthwick (1977) and Orvis and associates (1981) have confirmed the existence of this critical few in the context of alcohol problems. In the latter study it was found that approximately 14 percent of the alcoholic population accounted for over 95 percent of the loss attributable to alcoholism (not including excessive drinking) among armed forces personnel. It is probable that the same principle of the critical few applies to some extent in the domain of mental health problems generally. This principle appears not to be a function only of symptom severity, the most impaired being the critical few. It suggests rather that even at the same level of symptom severity, some people will cope better than others. If we accept the existence of the critical few as a reality, it leads us to speculate that it is representatives of this group (defined on whatever dimensions) who are most likely to appear in EAPs, at least within the formal part of the framework of such programs. If this is the case, we would have to conclude that the relatively low penetration rates achieved by EAPs are disproportionately important since they succeed in identifying the worst loss-causing people in the work force. We must quickly note, however, that such a conclusion is based more upon deduction than empiricism: the field is badly in need of studies in which the psycho-social and job performance profiles of referrals are compared on the same dimensions with the profiles of the work forces from which they are drawn. Only in this way can we confirm whether the people most responsible for loss are actually being identified by EAPs.

Assuming for the sake of argument that it is indeed the critical few who find their way into EAPs (at least those with healthy formal components), the fact remains that many people are left unidentified and unhelped. While it may be true that these other members of the population at risk do not represent as great a source of loss to their employers, they can nonetheless be seen as depressing the overall productivity and probably the morale of the work force as a whole. At the very least, they are functioning below par. It is hard to imagine, for example, that the kinds of people described in chapter 3 who were not in the highest (clinical range) distress categories but who were nonetheless acutely or chronically troubled could have been giving their best, assailed as they were on a daily basis by a multitude of unpleasant and distracting symptoms of anxiety, depression and other disorders.

Taken as a whole (with the inevitable risks attendant upon generalization), the "moderate-risk group," as we may call these people with subclinical distress levels, are people who have become to one degree or another frayed around the edges, worn out, ground down, burned out. They include the chronically weary, the anxious, the depressed, the overweight, the super-stressed, the compulsive smoker, the excessive (but not alcoholic) drinker, the unfit, unrested, unstimulated, and unchallenged. That they represent losses to themselves, their families, and their employers as a result of these conditions has been documented in terms of premature morbidity, mortality, decreased job satisfaction, and productivity (see, for example, Selye 1976; House 1974; McQueen and Siegrist 1982; and chapters 2, 9 of this book). However, members of this troubled group vary a great deal in the severity of their difficulties with some being close to candidacy for EAPs while others are not. In either event, EAPs, as we have shown, typically do not reach people in this category. It is possible that on solely a *cost-recovery* basis, it would not even be wise to attempt contact with the majority of this population through EAPs given the usually expensive nature of this type of intervention. It has been noted by Swint (1982) and Schramm (1982) that the cost–benefit ratio of EAPs tends to skew toward cost as programs push further toward earlier identification. This assumes, however, that EAPs remain the creatures we know: programs that are characterized by the provision of services by professional care givers usually on a one-to-one basis. In the case of alcohol-related problems, we may add that this care is often provided in very expensive residential settings in spite of evidence that suggests that less heroic forms of intervention would be equally effective (Edwards et al. 1977; Orford, Oppenheimer, and Edwards 1976; Orvis et al. 1981).

In spite of these reservations, it appears that more often than not EAPs do achieve cost recovery or better in the few instances in which this has been studied (for example, Foote et al. 1978).

Cost-recovery considerations apart, many if not most people in the moderate-risk group will probably not perceive themselves as being sufficiently troubled to warrant therapeutic intervention of the kind adumbrated in the foregoing. It becomes necessary, therefore, to consider either the modification of EAPs to include credible educational and training components for employees or the introduction of new or redirected interventions from another perspective altogether—that is, the health promotion perspective. Both options may indeed be feasible. The last part of this chapter is given over to a discussion of the potentially delicate interplay between EAPs and HPPs. In raising the issue that EAPs and HPPs may coexist not only peacefully but synergistically, the present writers are aware that a sharp debate is already underway between advocates of some kind of marriage between the two forms of intervention and those who fear that such a union would severely undercut the effectiveness of EAPs. We do not presume to argue that these fears are entirely without foundation. Roman (1984),for example, has suggested that EAP professionals may lose their identity and credibility if they are perceived or present themselves as being all things to all men and women. This is indeed a scenario to guard against and yet the vision of such a tragedy (for such it would be, given the great benefits that often accrue from EAPs) is predicated upon the idea that EAPs and HPPs would be *integrated* rather than *coordinated*. While we shall elaborate upon this as we proceed, our opening argument is that the interventions should be coordinated, even to the extent of joint policy statements, but not integrated, that is, EAP and HPP professionals would retain their identities but in the context of a more cooperative division of labor (see Miller, Shain, and Golaszewski 1985 for a fuller treatment of this issue).

One of the factors that will increasingly influence the cost-recovery picture in EAPs is the use of off-site, third-party service providers, although the use of such services appears to be as often considered within a purchase of service model. Even so, it is important to understand which program components are being juggled in order to arrive at different costs in such a model, since ultimately the issues of service delivery standards and value for money arise.

When third-party service vendors state or estimate a dollar figure for the provision of their services, there is often an ambiguity about what will actually be provided. Among those who charge on a "capitation" basis (that is, so much for every member of the work force covered), there is an enormous range of costs, many known to the authors falling within a span of $10.00 to $45.00 per capita. It appears that *at least* the following factors are involved in explaining this variation. Most of them concern the scope, intensity and quality of service.

Factor 1. The size of the estimated case flow from a given client organization. For example, if the allocated staff time is based on a projected 2 percent of the work force showing up for help in the first year, and it is believed that an average of five hours will be required per case at X dollars per hour, the pro-rated per capita cost will obviously be less than if the estimated case flow is 5 percent in the first year. Of course, vendors also need to allow themselves a margin, particularly when they may have to absorb the cost of losing a contract in cases where they are serving a number of different client organizations and no subsidization is available.

Factor 2. The amount of direct service time the vendor proposes to spend or actually reports having spent on cases, on the average.

1. Are "assessment" and "referral" functions that are performed by the vendor? If so, what procedures are used? Are they standard? How long do they take?

2. Are primary care and follow-up functions built into the service? If so, what standard is being used by the vendor? For example, what frequency of follow-up is proposed, on the average?

3. How many (what proportion of) cases are referred to other agencies in the community? Obviously, it makes an enormous difference if the vendor is simply "pipelining" referrals after a quick assessment as opposed to handling a large percentage on a casework basis. The service purchaser may be paying only $10.00 per capita but be receiving less than value if cases are simply being pipelined. Correspondingly, the purchaser may obtain a raft of services from another vendor for $20.00 per capita, involving intensive casework, follow-up, training, education, and program maintenance.

   Of course, when the service provider is acting only or principally as a pipeline, the odds are raised that further costs to the employer will arise in the form of charges from other treatment agencies to which the vendor refers cases. The nature and extent of these charges depend on the insurance coverage for treatment that the employer has in effect at the time. Although some charges are offset by insurance in the short run, the longer-term impact on premiums can be undesirable. Thus, a sensitivity to the appropriate use of outpatient versus inpatient treatment and to optimal use of short-term counseling is required on the part of both service vendors and purchasers if costs are to be contained.

   To put the matter of "pipelining" into some perspective, it is worth noting as a hypothetical example that if 1 percent of a 450,000 strong work force in 500 organizations with EAPs were to be referred into the community treatment network in one year, this would result in a burden of 4500 cases on the health care delivery system. If only 60 percent of the 500 organizations generated cases at this rate, it would still be 2700 cases. It is easy to see that without careful case management, the indiscriminate dumping of this many cases could create havoc in the care delivery system.

4. Related to the above, what is the average duration of a direct (face-to-face) contact among those cases that are eventually referred and among those that are not—that is, how many hours are spent per case?

5. What is the breakdown of these cases according to whether they are seen individually, in a group, in a couples situation, a family situation or with "significant others"? Group practice may not always be adopted as the most effective measure—it may have more to do with cutting costs. On the other hand, family work is time consuming even though it may be recorded as only one "case" in some instances.

6. How many cases are eventually seen on a direct (face-to-face) basis?

Factor 3. The range of services beyond assessment, referral, and direct face-to-face counseling provided by the vendor.

1. How many "consultations" are performed or are expected to be performed. This service refers to giving advice to managers and others who are having difficulty with an employee. This level of service may be encouraged by the vendor in which case a brisk trade can be anticipated. Some estimate of the time spent at this level could then be expected. As contracts mature one might expect a reasonably stable ratio between direct counseling and consultations to emerge.

2. Is there a 24-hour hotline service? If so, is it used?

3. Does the vendor undertake training of supervisors, union officials, committee members, and management with regard to policy and program utilization? If so, how many hours are devoted to this and over what time period?

4. Does the vendor undertake education of the work force about policy and program utilization and/or about specific topics related to utilization—for example, giving information about health risk factors (alcohol, drug, tobacco use, etc.)?

5. Are any other policy/program maintenance functions performed by the vendor? If so, what and when?

These levels of intervention are likely to be highly interactive: if the vendor does a lot of training and education, more cases can be expected to emerge. The clearer the pathways to care are made, through policy/program maintenance, the more likely it is that people will use them. This may result in *higher* costs in subsequent years: as noted later, this may not always be matched by corresponding benefits to the employer.

Factor 4. The qualifications and expertise of the staff. Usually there will be a relationship between this factor and service cost. Furthermore, the kinds of problems that the service provider deals with will be influenced by the actual and perceived qualifications and expertise of the staff.

As pointed out earlier, the idea of benefits in the purchase of service model is somewhat broader than in the cost recovery approach. Even so, it will be as well to keep in mind the fact that an optimum cost–benefit ratio is likely to be achieved by paying attention to the match between client needs and type of intervention. For instance, the U.S. Air Force recently demonstrated the difference in the cost–benefit ratio when the relative effectiveness of inpatient alcoholism treatment for moderately dependent problem drinkers was compared with outpatient treatment for the same group. To offset the cost of *inpatient* treatment an enlisted man would have to provide another four years of problem-free service while the same man would only have to serve twenty-one months to offset the cost of *outpatient* treatment (Orvis et al. 1981). Given that for this *moderately* dependent group both treatments were equally effective, the more favorable cost–benefit ratio would be achieved by the use of outpatient treatment.

We should reemphasize that it is essential to identify whose criteria are being used to define "benefits" and over what period of time these are expected to accrue and last. In this regard, unions, management, and program staff are often not at one, requiring specific negotiations about program expectations, acceptable costs, and predicted benefits.

As noted earlier, if we look at the published literature on cost–benefit in EAP we come away with the view that cost recovery and the kind of accounting involved in purchase of service approaches are paramount in the minds of employers. However, the results of three studies fly in the face of these assumptions. These were done by Roman (Executive Caravan studies, 1980), by Sapolsky and associates (Fortune 500 study, 1981) and by Groeneveld and associates (Canadian National Railway study 1984). In most cases there was little concern among executives about cost control in established programs. In Roman's studies the principal concern was with effectiveness of rehabilitation, while in Sapolsky and associates' it was a desire not to rock the boat by meddling with negotiated benefits packages. In the Groeneveld and associates study (1984), as many executives said in interviews that the *dual* objectives of their EAP were effective rehabilitation and cost control as said that saving money was the prime object. Of course, there are bargains to be had and we are not trying to suggest that executives are not mindful of them. For example, some large hospitals will discount health care costs to major corporate clients while some companies prefer self-insurance, which allows them the freedom to invest their own funds. However, these arrangements are made to optimize cost–benefit ratios in health care provision rather than necessarily to produce a black figure on the bottom line.

HPPs and Cost–Benefit

As noted in chapter 1, HPPs refer to a very wide range of interventions dealing with such matters as cardiovascular health, nutrition, weight loss, hypertension control, stress management, smoking cessation, fitness, life-style reappraisal and overhaul. The method of approach to these areas varies according to the diverse backgrounds of the people who ply their trade in the health promotion domain. Some of them are members of recognized professions (physicians, physiologists, kinesiologists, psychologists, health educators), while others, although not members of recognized professional groups, seek to establish reputations based on results. The latter group in a sense parallels the early EAP professionals who needed, and in some cases still need to establish credibility with employers, workers, and their unions. Partly because of the differences in professional background and partly because of variations in funding and organizational commitment, HPPs run the gamut in terms of efficiency and effectiveness. A useful overview of program types and commentary upon their effectiveness (in so far as it is known) has been produced by Parkinson and associates (1982). Underlying most programs is the desire to mobilize participants' personal resources in an effort to reinforce their strengths and bring about a renewed ability to control personal (and sometimes familial) health. (Evaluations of programs in this genre appear in chapters 5 and 6 of this book.)

Given the variety of these programs, a corresponding diversity of evaluative criteria and associated cost–benefit considerations have emerged in relation to health promotion in the workplace. However, as with EAPs, there are still two general dimensions of program performance from which this diversity of specific application stems; namely, *coverage* and *effectiveness*.

In HPPs, coverage is usually discussed in terms of *participation rates* and *program adherence rates,* while effectiveness is considered in terms of *success rates,* which in turn have both economic and public health aspects. Expectations for program success—whether it be for smoking cessation, weight reduction, or stress management—include: increased productivity; reduced turnover of staff and longer worklife; lower authorized and unauthorized absences; higher morale and job satisfaction; higher employee predictability; better corporate image for recruitment; less sick benefits and long-term disability (LTD) disbursements; fewer accidents and workers' compensation (WCB) claims; and lower health insurance premiums.

Few precise studies have been done in which gains on even the harder cost lines just mentioned have been quantified in relation to HPPs. The best evidence seems to be in the field of smoking cessation and hypertension control (Fielding 1982; Kristein 1982) and here again success depends upon a wide variety of factors associated with program design, characteristics of the host organization and its surrounding community, and the employees themselves. Fitness and exercise programs have shown gains in cost recovery terms in one study (Cox, Shephard, and Corey 1981), but this research needs replication in other settings before firm conclusions can be drawn. The efficiency of this program (Canada Life Project) was a function of very high participation and adherence rates. Absenteeism and turnover in the participating group decreased significantly in comparison with a control group. Productivity was not measured, but it has been argued that very small increases in production quality or quantity (in the neighborhood of 1 percent) would be required to offset program costs (Faust and Vilnius 1983, 527).

In so far as HPPs are aimed at the control of risk factors such as smoking, high blood pressure, cholesterol levels, obesity, high stress, excessive alcohol use, poor dietary and exercise habits, and so on, it is as yet unclear whether it is necessary to successfully achieve this control in order for reductions to be made in absenteeism and turnover. It has been suggested that the fellowship of shared activities and learning about health may be just as important as control of the risk factors themselves (Cox, Shephard, and Corey 1981).

Stress management programs are in themselves a major class of interventions with corresponding internal variety. Some of them have been shown to reduce the physiological and psychological manifestations of stress. Perhaps because of the difficulty involved, it remains to be demonstrated that any of them are effective in preventing premature morbidity and mortality (see Schwartz 1982, for a review). Chapter 6 in this book is devoted to a description of two stress management program evaluations. In both cases improvements were shown on the major indices of distress used in those studies. Given the relationship between high anxiety and poor self-efficacy noted there, it is almost certain that employers will benefit from any program that brings about reductions in experienced anxiety or in the frequency of symptoms associated with a variety of other potentially debilitating conditions.

With regard to weight control, Faust and Vilnius (1983) showed that being excessively overweight is associated, particularly among males, with abnormally high absenteeism even when the effects of smoking and alcohol use were controlled. It is worth noting, however, that being *less* than optimum weight for height and build was also associated, though not as strongly, with higher absenteeism. In the same controlled study, called the "Go to Health Project," Faust and Vilnius showed that a comprehensive diet and weight modification program brought about significant declines in "aggregate absence hours" over a two-year period. It was in this context that these investigators put forward the view that a 1 percent net increase in productivity among those who took part in the project was all the employer needed to offset the costs of the program.

Cascio (1982) has shown that smoking represents a clearly quantifiable loss not only to employers but also to the smokers' co-workers as a result of second-hand smoke inhalation. Cascio computes excess costs due to smoking on a number of cost lines—absenteeism, medical care, premature morbidity and mortality, lost time on the job, property damage, depreciation and maintenance, high insurance for buildings, accidents, wage continuation, effects upon nonsmoking co-workers. These excess costs can amount to the equivalent of a sizable proportion of the average worker's wage. Cascio's estimates suggest that smokers can easily cost between 20 and 30 percent more to employ than nonsmokers on the cost lines just indicated.

In spite of some promising results, health promotion in the workplace tends to be bedeviled with a number of shortcomings, among which are: (1) low average coverage rates (participation/adherence); (2) a tendency not to differentiate audience needs and to adopt a "one size fits all" attitude when designing programs (Merwin and Northrop 1982); (3) a tendency to approach risk factors such as smoking, drinking to excess, drug use, lack of exercise, and so on, as though they were isolated from one another rather than interdependent parts of an integrated "system" of beliefs, attitudes, and behaviors (the "single issue" approach); (4) a tendency to be faddish; (5) a tendency to expect too much too soon for too little; and (6) a tendency to ignore the mental health aspects of wellness. This state of affairs will probably improve as consumers of these services not only recognize their value but become critical of how and by whom they are delivered.

The objection to exclusive use of a cost recovery model with HPPs is quite similar to the one raised in connection with EAPs. Attractive though the prospect of cost recovery may be, it is unlikely to be established one way or the other save in a handful of well-conducted, limited-situation, very expensive studies. Meanwhile, HPPs proliferate. As consumers of such services do their comparison shopping, it is often a question of perceived value for money—a relative, not absolute, definition of cost–benefit. This does not mean, though, that employers and unions are not interested in improvements to programing that will *optimize* the cost–benefit ratio. Particularly, they may be interested in the synergistic relationship that may be developed as a result of thinking through the relationship between HPPs and EAPs as they concern the health needs of the whole work force and maximization of productivity. In this comprehensive approach, needs and interventions are matched as closely as possible to avoid the wrong services being delivered to the wrong people.

Cost–Benefit Rationale for a Comprehensive Approach

If we refer to our projected comprehensive approach to the mental and physical health of workers as an "Employee Health and Assistance Program (EHAP)," what might be its characteristics and why would it offer a cost–benefit ratio superior to an EAP or HPP alone or superior even to the two programs together but not coordinated?

Characteristics

An EHAP would have both economic and public health goals that would translate into objectives at primary, secondary, and tertiary intervention levels. Within such levels, specific target groups would be identified through needs assessment, and appropriate interventions would be implemented to meet these needs (Shain and Boyle 1985). An example of such goals and objectives follows.

1. *Economic Goal:* to minimize the costs associated with deficits in the physical, mental, and social health and well-being of the work force.
2. *Public Health Goal:* to increase the physical, mental, and social health and well-being of the work force.

Translated into *objectives* at primary, secondary, and tertiary levels of intervention, some examples might be:

a. *Primary level:* to reinforce the intent and will of people in basically sound health to remain so through examination, reappraisal, modification, and improvement of life-style factors that influence wellness, vigor, and productivity. Discussions of interventions at this level appear in chapters 4 and 5.
b. *Secondary level:* to facilitate the ability of people with developing health problems to engage in self-corrective action by providing skill building and training programs aimed at full restoration of health, vigor, and productivity. Discussions of interventions at this level appear in chapter 6.
c. *Tertiary level:* to facilitate the recovery of people with developed health problems by providing formal and informal pathways to remedial services aimed at full or partial restoration of health, vigor, and productivity. This level of intervention is usually seen as the bailiwick of EAPs as currently known.

The particular form that such comprehensive programs would take would vary a great deal, depending on a number of factors. The actual shape of the approach taken in specific sites should depend upon careful needs analysis by management, employee representatives, and relevant program staff from all levels. Success, however defined, will depend upon a "fit" between the needs of workers, organizational characteristics, and program characteristics. In that sense, the solution must creatively fit the *culture* of the work organization.

The Likely Benefits of an EHAP Relative to Costs

The core of the cost–benefit rationale for an EHAP is that potentially at least, the health needs of the whole work force are covered or considered. The typical problem to date with both EAPs and HPPs has been that they each deal with extremes: as figure 7–1 suggests, the typical target group (or at least the group that is in fact reached) in EAPs is the *"walking wounded."* Members of this group have typically well-advanced problems that affect job performance. The employer's response is largely oriented toward individual treatment by trained professionals. HPPs, on the other hand, tend to reach only the *"conspicuously well"* (in the case of fitness and exercise) or small proportions of highly selected target groups (smokers; highly stressed, over-weight people). Exceptions to this state of affairs exist in both types of pro-grams, as indicated by the broken lines on figure 7–1.

The *moderate* risk group, however, is typically untouched. Nonetheless, it contains, as suggested earlier, people who are experiencing physical and mental health problems that, if not resolved, will probably get worse, thus making them potential candidates for EAPs; people who are suffering acute situational difficulties that depress their well-being and productivity; and people in states of chronically mediocre health who could be feeling and per-forming much better. The condition of this group is associated in some im-portant ways with personal life-style practices that put them at higher risk for a variety of illnesses and diseases. Prominent among these practices are those associated with excesses or deficiencies in eating, sleeping, working, traveling, smoking, exercising, drinking, and obtaining respite from everyday troubles and worries.

There is an accumulating body of evidence which says that the groups just described are a significant source of loss to employers because of the net impact of relatively low-level problems on daily productivity, annual sickness and absenteeism rates, and general employee morale (for example, Faust and Vilnius 1983; Cox, Shephard, and Corey 1981; Cascio 1982). Even accident rates are affected by the existence of a large group of people in the workplace who are functioning below par (Mannello and Seaman 1979). One of the great potential advantages of the EHAP approach is that it deals with the organizational losses attributable to the moderate-risk group.

A second potential advantage is the optimal use of available resources deriving from a comprehensive approach. In figure 7–2, both EAPS and HPPs are shown as penetrating the moderate-risk group, but at considerably dif-ferent costs. This is a hypothetical projection, of course, and the cost projec-tions will vary enormously depending on the characteristics of the program and the characteristics of the work force in question.

PROGRAM TYPE

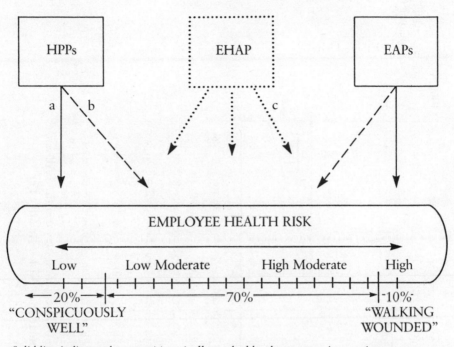

^aSolid line indicates the group(s) *typically* reached by the program in question.

^bBroken line indicates the group(s) *occasionally* reached by the program in question.

^cDotted line indicates the *projected* reach of the new program type (EHAP).

Figure 7–1. Typical and Projected Penetration of the Work Force by EAP/HPP.

In the projection shown in figure 7–2, HPP costs tend to escalate less rapidly than EAP costs when addressing the needs of the moderate-risk group. However, such a comparison assumes (1) that the same people in the moderate-risk group would seek help through an EAP as would attend a HPP and, (2) that the two types of program would be equally beneficial to participants. Neither assumption has been validated. However, the possible additional effectiveness of EAP-based clinical interventions with moderate-risk people is offset by additional costs with probably little chance of recovering them, while the lower effectiveness of usually group-oriented, skill-building

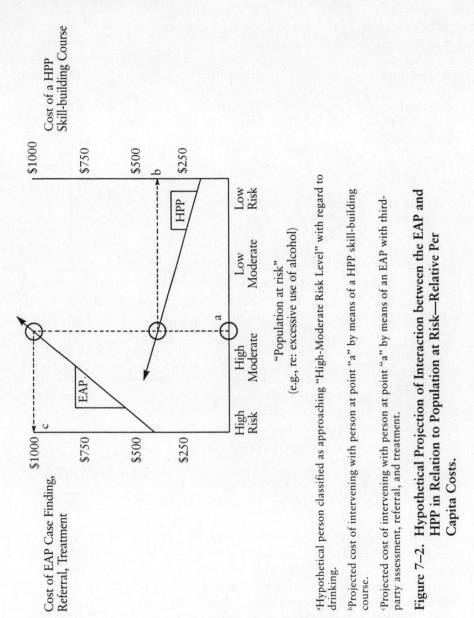

"Population at risk"
(e.g., re: excessive use of alcohol)

[a]Hypothetical person classified as approaching "High-Moderate Risk Level" with regard to drinking.

[b]Projected cost of intervening with person at point "a" by means of a HPP skill-building course.

[c]Projected cost of intervening with person at point "a" by means of an EAP with third-party assessment, referral, and treatment.

Figure 7–2. Hypothetical Projection of Interaction between the EAP and HPP in Relation to Population at Risk—Relative Per Capita Costs.

HPP approaches is offset by proportionately lower costs. The escalation of costs in EAP as it progresses toward, in its terminology, "earlier identification" is due to the added resources required for supervisor training, employee education, and case finding as the candidates become less obviously troubled. Some of these costs may be in turn offset by the increased voluntary use of programs and by the provision of less intensive forms of treatment for problems, for example, outpatient care, day treatment, weekend programs, and so on. Thus, the cost–benefit ratio is really a sliding scale that can be roughly captured by the chart but that in practice has to be used by concerned parties on the basis of *trade-offs* between desired outcomes and available resources. The term trade-off, it may be recognized, is one that belongs to the language of cost–benefit at any level except that of strict cost recovery, although it does not preclude attempts at comparing the costs of different methods of service provision.

Further reference to figure 7–2 will show that on a strict cost-recovery basis, EAPs and HPPs appear to work most efficiently at the extremes of the health continuum: that is, with the "walking wounded" and the "conspicuously well." However, in terms of *benefit* to the organization, the gains are not distributed in the same way. In the EAP, as noted earlier, the "critical few," who roughly correspond to the "walking wounded," are a small subgroup of very costly problem employees who tend to get identified and referred at an early stage of program development. Since the *cost line* for the EAP is also very roughly a *time line,* one can see that the greatest efficiency for the EAP tends to be in the start-up period when the critical few are being identified, although it is also true that start-up costs tend to be high and are amortized over successive years. In the HPP the situation is different. The gains that accrue from reaching the conspicuously well in early-stage program development are less than those that start to accrue once there is penetration of the moderate-risk group. So, if the cost line is also the time line with the HPP, greater gains can be expected to accrue *later* in program development, although it is not clear yet whether these benefits offset costs in strict recovery terms.

However, all this needs to be considered in the context of two other benefits of the comprehensive approach. The first is that there will probably be a tendency for *cross-referral* to develop between the EAP and HPP particularly where there is cross-fertilization of program staff; for example, an EAP publicity component can be built into HPPs at the basic educational level. HPPs at the training and skill-building level can offer referral pathways into more intensive EAP interventions, while clients of EAPs can be referred for auxiliary care into HPP skill-building programs such as stress management, life-style reappraisal, smoking cessation, and so forth. In fact, as figure 7–3 shows, there should always be an EAP component in HPPs and vice versa.

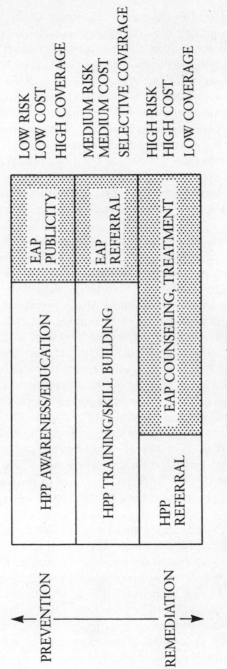

Figure 7–3. Coordination and Synergy of the HPP and EAP

The final advantage of the EHAP is that it requires comprehensive planning. For this to occur, there has to be a process within organizations whereby all concerned parties—management, staff representatives, health professionals and the like—come together to discuss the health needs of the total work force. Such a process in and of itself is part of a change in corporate culture and represents a shift toward the ecological model of cost–benefit referred to earlier. EHAPs can potentially provide a barometer of organizational functioning via the feedback received by program staff from participants and clients. This feedback, delivered to managers at the most senior levels, can serve as a measure of the health of the organization. A report of how this approach was used successfully at Johnson and Johnson in their "Live for Life" program appears in Wilbur and Garner (1984).

Another practical model for the development of EHAPs has been described by Carey (1985). Carey outlines how the Human Resources Council was developed at the New England Telephone Company. This council comprises members from all key human-resource departments in the organization, embracing health promotion (including medical department), EAP, safety, personnel, and benefits representatives. According to initial reports, this council functions well in developing an overall policy toward the health of the work force as a whole, while preserving the autonomy of individual divisions and branches. It appears to be a coordinative rather than an integrative model.

In spite of its theoretical appeal, the move toward EHAPs will probably be difficult. We can expect in the short run a great deal of debate and acrimony between programmers who currently ply their trades at opposite ends of the health continuum, namely EAP and HPP. Philosophical and practical differences will have to be identified and where possible ironed out (see Miller, Shain, and Golaszewski 1985). Competition for scarce resources will have to be minimized. However, the potential benefits of a comprehensive approach are enormous. An EHAP is more than the sum of an EAP and HPP. It represents a move toward a greater and more proactive concern about the value of human life and about working people as not only producers of goods and services but as robust members of society. The benefits of such approaches, then, will accrue not only to the workplace but to society as a whole.

This discussion has proceeded with little attention paid to the socio-technical structure of the workplace and how it affects the nature and quality of programs that emerge within it. In so far as the issue has been raised, it was suggested that comprehensive health planning could give rise to benign organizational processes that might lead to better overall working conditions. There is a darker side to this picture, which has been intimated in Chapter 2 and will be further explored in Chapter 8. Some of the concerns are that programmatic solutions may be seen as bandaids in situations that require structural overhaul and that the "health culture" may carry with it a pressure toward conformity that may be seen as the product of the worst kind of patronization or even as a sort of tyranny. While these issues are taken up elsewhere in this book, it is important to weigh the arguments presented in this chapter in the context of these structural considerations.

8
The Impact of the Work Environment upon Mental Health: Implications for the EAP and HPP

Throughout this book we have made reference to the fact that environment, in the broadest sense of socioeconomic, political, and cultural conditions, has far-reaching effects upon individual beliefs, attitudes, values, and behavior. These effects are nowhere more marked than in the case of work organizations. The purpose of this chapter is to examine the impact of work environments upon mental health with a view to deriving implications for employee assistance and health promotion professionals. We believe that these professionals have an important role to play in the identification and even modification of problems in the work environment itself.

Many EAP and HPP practitioners would probably accept as a truism that the nature and organization of work affect mental as well as physical health. However, this is often a rather superficial acceptance of an apparently bland proposition. Just as often, practitioners go about their daily work with only a dimly formulated perception and a vaguely felt unease concerning the importance of workplace conditions, and yet these conditions exert an extremely significant influence upon the success of their own endeavors. In so far as this influence is recognized, the ability to modify it tends to be seen as falling outside the sphere of EAP and HPP competence. Professionals in these areas (particularly HPP) tend not to see themselves as organizational engineers and do not feel that others see them this way either. The result, unfortunately, is that programs are frequently run with an apparent unawareness of how their outcomes are likely to be affected by the environmental conditions in which they emerge. The temptation for programmers who are frustrated or intimidated by the seeming enormity of the job involved in modifying the structure of the workplace is to simply run their programs as though their success were not influenced by such factors at all. Regrettably, the result tends to be that program participants are subtly or not so subtly led to perceive that the preservation or restoration of health is entirely their responsibility.

The idea that EAP and HPP practitioners should take on yet another role, and a very difficult one at that, may strike many readers at first glance as an absurdity. In response, we argue that a fuller understanding of what is meant by the impact of environment may suggest methods of attack that are not so alien to the roles of the professions as currently understood. That is, the proper formulation of a problem may go a long way toward its solution.

The diversification of roles for EAP and HPP practitioners is clearly a contentious issue that cannot be resolved or even fully described in this book. However, we can hardly fail to mention that the professional status and credibility of these practitioners is what appears to be at stake in this matter. Roman (1981b, 1984) has argued, for example, that EAP staff and consultants should stay away from areas in which they are not, or are not perceived to be, competent. Thus, he says, they should not involve themselves in health promotion. Francek (1985), on the other hand, argues that occupational social workers should be involved in the process of organizational change, although he does not necessarily endorse their involvement in health promotion. Further, Francek is speaking mainly about occupational social work, not about the role of psychologists, psychiatrists, nurses, physicians, or others who might ply their trades in the workplace.

Perspectives on Environment and Mental Health

Several proxy measures of "mental health" have been used in studies generally related to our area of concern. As we review the various approaches to the organizational determinants of mental health in the workplace, it will become apparent that we are dealing with the politics of research as much as with reality. Otherwise stated, we are dealing with multiple realities, each constructed by investigators or commentators from the perspective of their own world views and projected into their research paradigms, methods, and conclusions.

The Stress and Strain Approach

One major thread in the fabric of research on how environment affects mental health consists of studies related to stress and strain. The terms stress, stressor, strain, and distress unfortunately have been used with virtual abandon in the literature. However, some authors do try to distinguish carefully among the various terms, usually on the basis of locus. For example, McLean (1974) indicates that strain is experienced by the person, whereas a stressor is a factor in the environment. (In this chapter, unless otherwise specified, "stress" is used synonymously with "stressor" to refer to the environmental factor.)

Most discussions of mental health problems in the workplace include some reference to strain, pressure, tension, or related conditions. Usually the connection is postulated to be that some aspects of the work environment operate as stressors that lead to health problems, both mental and physical. Margolis and Kroes (1973) identify five dimensions of job-related strain (here referring to what the individual experiences):

1. Short-term subjective states related to specific job stressors (for example, fear when the boss criticizes you);

2. Chronic psychological responses (for example, chronic depression, fatigue, alienation);

3. Transient clinical psychological changes;

4. Physical health status (for example, heart disease, asthma);

5. Work performance decline.

Such studies also tend to emphasize the importance of "mediators." Mediators can create a large difference between one person and another in the impact of objectively similar environmental factors. These mediators fall into two main classes. First, there are the characteristics and resources of the individual in the sense of needs, expectations, and personality on the one hand, and learned methods of coping with difficulties on the other. Secondly, there are potential "cushioning" features of the environment, such as supportive work groups, friends, relatives, community, church, club affiliations, and so forth (Rabkin and Struening 1976).

One of the best-known studies on the mental health of workers was done by Kornhauser (1965), who defined mental health as "those behaviors, attitudes, perceptions, and feelings that determine a worker's overall level of personal effectiveness, success, happiness, and excellence of functioning as a person" (p. 11) and indicated the importance of individual mediators by stressing that crucial to mental health was "the development and retention of goals that are neither too high nor too low to permit realistic, successful maintenance of belief in one's self as a worthy, effective human being, a useful and respected member of society" (p. 14). His interviews with Detroit auto workers revealed that mental health varied consistently with job level, with workers on skilled jobs having the highest mental health scores and workers on repetitive jobs having the lowest. Two general job characteristics seemed to be most relevant to the difference in mental health between occupational groups: first and foremost was "the opportunity the work offers—or fails to offer—for use of the worker's abilities and for associated feelings of interest, sense of accomplishment, personal growth, and self-respect" (p. 263); the second characteristic concerned feelings with respect to income and financial stress at upper and lower skill levels. Men with good mental health were more likely to be satisfied with their job; feelings about the job were identified as the intervening link between the job and its impact on mental health. Kornhauser placed part of the blame for impaired mental health and dissatisfaction on a poor fit between goals and aspirations on the one hand and opportunities to fulfill them on the other. This congruence of goals and reality is a subject to which we shall return shortly.

Kahn and associates (1964) explained organizational stressors in terms of role theory. From an intensive study of 53 "focal people" drawn from six industrial locations and of the "role senders" identified by these people, and from a national survey of 725 employed workers, they found cause to implicate "contradictory role expectations" in "intensified internal conflicts, increased tension associated with various aspects of the job, reduced satisfaction with the job and its various components, and decreased confidence in supervisors and in the organization as a whole" (p. 70–71). Similarly, "ambiguity" was related to increased tension and reduced trust in associates. However, ambiguity about tasks to be done seemed to create job dissatisfaction and feelings of futility, while ambiguity about one's evaluation by others appeared to weaken the person's relations with them and his or her own self-confidence. Some of the major organizational sources of conflict and ambiguity were identified as the requirement to cross organizational boundaries (for instance, contacts outside the company), the requirement to produce creative solutions to nonroutine problems while at the same time dealing with routine tasks and with people who were status-quo-minded, and the requirement to be responsible for the work of others.

Another study examined perceived job pressure among managers and workers in a manufacturing company (Buck 1972). Job pressure was defined as "the resultant psychological state of the individual when he perceives that (1) conflicting forces and incompatible demands are being made upon him in connection with his work; (2) at least one of the forces or demands is an induced one; and (3) the forces are recurrent or stable over time" (p. 173). Among managers, 78.5 percent of the variance in perceived job pressure was explained by environmental factors alone; this figure increased to 86.4 percent when personal variables were added. The corresponding percentages for workers were 53.5 and 71.4 percent. Perceived job pressure was related to decreased overall job satisfaction and sufficiently to decreased mental health for the author to comment that "there is a cost to the individual who is working under pressure in terms of his own outlook on life and the fullness of his existence" (p. 167).

Margolis, Kroes, and Quinn (1974) interviewed a large number of employed individuals to try to establish a relationship between various stressors in the work environment (role ambiguity, underutilization of employee, overload, resource inadequacy, insecurity about continued employment, lack of participation in job-related decisions) and indicators of strain in the workers (overall physical health, escapist drinking, depressed mood, self-esteem, life and job satisfaction, motivation to work, intention to leave job, frequency of suggestions to employer, absenteeism). They found that overall job stress was significantly associated with all the strain measures except frequency of suggestions to employer, and especially strongly with job satisfaction and work motivation. Nonparticipation in making decisions related to work was the stressor variable that had the highest correlation with eight of the strain measures, while the strain variable of self-esteem correlated highly significantly and negatively with all the stressors except overload.

In a Swedish survey cited by Wahlund and Nerell (1976), the immediate job stressors were identified as duties that were excessively wide-ranging, too advanced, or insufficiently defined; teamwork in the case of older people who had a slower pace or different values from their younger co-workers; concern about future reorganization; contact with the public with respect to controversial social questions; and exacting demands in the area of cooperation with superiors, subordinates, and colleagues. Wahlund and Nerell's (1976) own survey of approximately 12,000 white-collar Swedish workers revealed experienced mental strain to be related to a lack of liberty in the job (inability to move around or take breaks freely); a feeling that others, especially those outside the workplace, had excessive control in the planning and execution of the work; overstimulation (too much responsibility, demands for independent initiatives and decision making, concentration, and alertness); excessive work load or insufficient time; expectations from outside the workplace (such as from customers, students, patients, or the public); and less significantly, lack of help from the boss or from colleagues.

After factor-analyzing scores on questionnaires completed by 133 male employees of a national accounting and management services firm, Cummings and DeCotiis (1973) found three major organizational correlates of both psychological and physical strain: perceived organizational support and clarity, perceived organizational objectivity and rationality, and perceived administrative rationality.

Stress research as related to the workplace is not by any means "value-free." As we suggested earlier, politics can enter into this area very easily. A good example of this (referred to also in chapter 2) is the dispute between the Air Traffic Controllers and the Federal Aviation Authority (FAA) in the United States in 1981. It has been observed that air traffic controllers who are responsible for large numbers of lives and who are in voice contact with the airplanes that contain these lives, have significantly higher rates of hypertension, peptic ulcer, and diet-controllable diabetes than second-class airmen (Cobb 1973). However, when these and similar facts were used by the controllers' association in an effort to improve working conditions, the FAA took the view (in resisting such changes) that the adverse reactions to stressful situations that were being reported by the controllers resulted from their inability to cope. In other words, it was their problem. The controllers, on the other hand, saw strain and related health disorders as products of work environments that were technically and socially engineered so as to maximize the severity of these adverse conditions (Tesh 1984).

Stress and strain are concepts that do not in and of themselves suggest the source of responsibility for their creation. It is often difficult to determine the implications of stress research in which both stressors and personal mediators are identified, since the latter may or may not be perceived as falling within the individual's control.

Occupational Alcoholism Studies

The excessive use of alcohol by employed people can be studied from the perspective of how both environmental and individual characteristics influence this behavior. In fact, research on this topic can be classified along a spectrum ranging from studies predicated on the view that the individual brings drinking problems to the workplace, to studies that see even excessive consumption as a behavior largely conditioned by or responsive to the nature and organization of work. It can be readily appreciated that this situation is analogous to the one just described in relation to strain.

As an example of the individual predisposition approach we may cite Hitz (1973), who suggests that heavy drinkers may select occupations where drinking is accepted, encouraged, or at least not interfered with, and indeed that certain jobs may recruit from demographic areas that have a higher rate of problem drinking. On the other hand, cliques and after-work socialization, with drinking as an integral aspect of the activities, may be the norm in some jobs and the new employee may fall or be pressured into the pattern. Hitz attaches little credence to the view that work may "drive people to drink," although she admits that it could happen.

Much of the earlier work done on variations in drinking practices in fact focused on the *occupations* of employed people. Explanations were sought for variations between occupations, on the supposition that certain kinds of work attracted certain kinds of people and that part of the appeal of some jobs was the availability of alcohol or condonement of its (excessive) use. Alternatively, some occupations were seen to involve intrinsic stressors (such as isolation, pressure) that contributed to excessive "relief" drinking. A whole series of studies operated upon one or both of these assumptions (Plant 1977).

Parker and Brody (1982) were instrumental in providing a balance to the view that drinking in the workplace could best be understood exclusively by an examination of differences *between* occupations. Their research allows us to see that variations in the prevalence of drinking *within* occupations are instructive with regard to the different conditions of work that contribute to such variations. In taking this approach the investigators were influenced by the work of Roman and Trice (1970), who in effect identified environmental factors that vary within as well as between occupations. The factors to which they referred bear a close resemblance to those reviewed earlier in this chapter: clarity of production goals; amount of feedback on performance; flexibility and structure of hours; role clarity; degree of accountability; degree of formal and informal control over deviant behavior. The importance of Roman and Trice's contribution lies perhaps not so much in the identification of these factors as in the generation of hypotheses about how they relate to a diversification of drinking practices. Parker and Brody developed measures of Roman and Trice's environmental factors in their study of employed men and women in the Greater Detroit area. Their object was to relate these measures to job problems and to drinking behaviors, including "loss of control" drinking.

Parker and Brody themselves did not take much interest in the analysis of work conditions that might be related specifically to the differences in drinking patterns according to occupation. Instead, they proceeded to look at their sample of 564 women and 484 men according to factors that must be presumed to *cut across* different occupations—for example, amount of supervision, job visibility, job structure and complexity, job stress, and so on. We are left to deduce the relationship between the operation of these general factors and the way in which specific occupations are structured in different workplaces.

Job visibility was operationalized as two variables: "job routinization" and "degree of supervision." The former did not in itself discriminate among people with differing drinking patterns although, as we shall see, it did so when mediated by the influence of stress. Degree of supervision, on the other hand, did discriminate between problem and nonproblem drinkers before other socioeconomic variables and job complexity were controlled. However, it discriminated in a direction that appears to be contrary to the expectations of Roman and Trice: tight supervision seemed to be more associated with self-reported problem drinking than loose supervision. This relationship lost statistical significance when controls for the additional factors just mentioned were added. All this means, though, is that jobs that were tightly supervised were usually done by people of lower socioeconomic standing, at least where such jobs were simple and repetitive in nature. People in such occupations apparently experienced more drinking problems. These results, however, do not suggest that it was the socioeconomic characteristics of the workers that accounted for heavier (or more problematic) drinking, even though Parker and Brody seemed to imply that this was the case. Taken at face value, the data say only that drinking problems tended to be part of a syndrome that included low income, less education, being younger, being unmarried, and doing simple, mechanistic jobs.

Further analysis of the data by Parker and Brody involved the construction of three more variables: job complexity, time pressure, and job stress (which, according to the definition we are using here, really referred to job strain). It was found that lower levels of job complexity were associated with psychological dependence on alcohol and problem drinking in males but not in females. Time pressure in itself did not discriminate. Job stress was related to psychological dependence, loss of control, symptomatic drinking, and overall problem drinking in men, and to all these except symptomatic drinking in women. When job stress was used as the dependent variable in multiple regression, it was found that time pressure and routinization of work affected stress levels significantly among women, while time pressure and low job complexity significantly influenced stress among men. These findings appeared to establish job stress as a mediating variable between alcohol problems and structural conditions such as job complexity, routinization, and time pressure. Further, as the investigators claimed, the results located the development of problem drinking in the same context as other mental health problems, for example, those that are generally classed under the heading of alienation (Kohn 1976). In other words, drinking behavior may be usefully considered as an expression of more general psychological functioning as related to occupational experiences.

The Detroit study does not deal in any direct way with the difficult issue of self-selection raised earlier in this chapter. However, the data do provide a perspective on the matter that is worth drawing out. Central to this perspective is the concept of stress discussed briefly in the foregoing. In order to argue that people select occupations that are characterized by excessive drinking and that it is therefore the personality of the individual that accounts for such behavior, it is necessary, following Kohn and Schooler (1973), to maintain, first of all, that people have an accurate knowledge of the working conditions that they select; secondly that such choices are unconstrained by socioeconomic influences such as upbringing and education; and thirdly that a conscious selection of *all* the features of the chosen occupation is made. With regard to the latter, one would have to argue that even the stressful elements of a job are chosen. Parker and Brody appear to have shown that experienced job strain significantly mediates the impact of other social and technical aspects of work. High job strain is related to drinking, yet it seems hardly likely that most of us will consciously select occupations in which we know we are going to be miserable. Similarly, the other propositions about accurate knowledge of working conditions and freedom of choice seem equally improbable.

Another more powerful argument against the deterministic self-selection hypothesis is built upon the observation that job dissatisfaction, poor morale, and alienation are conditions which can be reversed to some extent through modification of the organizational environment. This reversal ought not to occur if the self-selection hypothesis is valid. According to this line of reasoning, disgruntled people or people with problems would leave their place of employment if conditions were altered because the environment no longer fitted their needs. This exodus does not appear to happen on any notable scale. (See the section on "Impact of Workplace Innovations on Mental Health," below.)

The dual influence of individual predisposition and environmental precipitation as they contribute to drinking problems is illustrated by Plant (1978). In this study 150 male, newly recruited manual workers in three brewing and distilling companies (considered high-risk jobs) near Edinburgh, were compared with 150 similar workers in two other firms in the same area (lower-risk jobs). The workers were interviewed during working hours on such topics as drinking habits, how drinking had changed since they started their present job, what had attracted them to this job, past employment, legal and health records. The results of the study indicated that both predispositions and job characteristics played a role in the greater alcohol consumption of brewery and distillary workers, that people inclined towards alcoholism were drawn towards these trades and that once employed in them, the social pressure to drink, the readily available alcohol as well as the hot, dusty work environment encouraged drinking on the job and in large amounts. Among those who did not select the jobs in the distillery or brewery because of alcohol availability, but rather because a job simply happened to be available, the same pressures to consume nonetheless contributed to their drinking more than they did before their employment in comparison with similar people who went to work in the other manufacturing companies studied.

The Alienation and Dissatisfaction Approaches

One of the most obvious manifestations of impaired mental health is chronic and severe worker dissatisfaction, sometimes also referred to as worker alienation. Alienation, however, has some special connotations of its own, due largely to the influence of Marx (1964). As used in its Marxian sense, alienation refers to a mental state that is born of performing tasks with which workers feel no identification or from which they gain no sense of personal worth or meaning: that is, doing work that does not satisfy intrinsic needs but that is done solely to satisfy other extrinsic needs. Marx postulated that intrinsic needs could be met only by workers having substantial control over the product and the process of their labor. Failing this, workers would be alienated.

The concept of alienation has been elaborated by sociologists in such a way that it is now usefully considered as an aggregate or composite condition made up of related but distinct mental states. These are identified by Seeman (1959) as powerlessness, self-estrangement, normlessness, isolation (or cultural estrangement), and meaninglessness. Kohn (1976) has attempted to show that the first three aspects of alienation are related primarily to the degree of worker control over the *process* of production (which he defines with respect to closeness of supervision, routinization, and substantive complexity) rather than to control over the *product* (which he measures mainly in terms of ownership and hierarchical position); that is, the important thing is *how* production is carried on rather than *what* is produced. However, he does allow that these relationships may not hold for samples substantially different in culture from his own. The sample in question was of 3,101 American men in civilian occupations.

The factors in the work environment that have been linked with low job satisfaction resemble the factors that have been linked with strain. Both appear to be associated with organizational and environmental obstacles that create a lack of fit between the worker's aspirations, goals, needs, and what he actually accomplishes. Herzberg (1967), an eminent investigator in this field, postulated a motivation-hygiene theory in which two separate groups of variables were responsible for job satisfaction and dissatisfaction. The satisfier factors or "motivators" seemed to encourage better performance and effort while the dissatisfiers or "hygiene" factors were mostly environmental characteristics that helped prevent job dissatisfaction with little effect on positive job attitudes. In his original study, two hundred engineers and accountants in Pittsburgh were asked two separate sets of questions about work events: one concerning events that led to exceptionally good feelings, the other concerning events that led to exceptionally bad feelings. Factors that emerged as related to job satisfaction were achievement, recognition for achievement, work itself, responsibility, and advancement. Factors related to job dissatisfaction were company policy and administration, supervision, salary, interpersonal relations, and working conditions.

A recent study by Eisenstat and Felner (1984) has provided further support for Herzberg's conceptualization of satisfiers and dissatisfiers as lying on separate dimensions. In this case, burn-out among various social agency staff who had direct client contact was investigated. The authors disaggregated the concept of burn-out, showing that the nature of work (resolution of client problems) and the way in which that work was organized (socio-technical and structural aspects) were two distinct sources of satisfaction or the lack of it. One might derive pleasure from the one and not from the other or from both or from neither. However, it seemed fairly clear that in terms of burn-out, one set of satisfactions could not compensate for the absence of the other.

Other attempts to disaggregate global expressions of overall job satisfaction have also tended to support the two-dimensional theory of Herzberg. Gruenberg (1980) has shown that intrinsic satisfactions (freedom to plan work, chance to learn or try new things, chance to use skills/abilities) correlate more highly with one another than they do with extrinsic satisfactions (pay, security, kind of people worked with, kind of workplace). Further, intrinsic satisfactions are the best predictors of overall job satisfaction. The variation in results from component measures of satisfaction, however, is far greater than in the case of global evaluations. For example, Gruenberg showed that in a sample of 792 American workers that cut across a range of occupations, 90 percent reported that they were satisfied with their jobs. However, more specific questions revealed that 40 percent were dissatisfied with their pay and an equal proportion considered that their chances to learn or try new things were no better than fair or poor. There was no evidence in this study that workers from lower-level occupations valued intrinsic rewards any less than their more exalted counterparts. However, there was some evidence that the relevance of these intrinsic rewards to people from lower-echelon jobs was a function of perceived availability; that is, if "meaningfulness" of work was seen as a desirable but unattainable goal, it tended not to be considered in evaluations of job satisfaction even though it may have been psychologically relevant.

The Job–Person Fit Approach

The concept of job–person "fit" refers, as its name suggests, to the degree of compatibility or congruence between the needs of the worker and the needs of the organization. This fit has been described in various ways. For example, Levinson and Weinbaum (1970) speak of the individual's concept of how he or she would like to be (ego ideal) and of his or her appraisal of how things really are (self-image). These writers argue that organizations have a positive impact on mental health when they provide opportunities for closing the gap between self-image and ego ideal. When organizations create conditions that widen this gap, however, they contribute to mental illness.

According to Van Harrison (1978), there are basically two types of fit between the person and the environment. First, there is the extent to which the person's skills and abilities match the requirements of the job. Secondly, there is the extent to which the job environment accommodates the individual's psychological needs. Poor person–environment fit is often hypothesized to lead to health strain, which can reveal itself, as we have seen already, in the form of psychological responses (job dissatisfaction, depression, low self-esteem), physiological responses (high blood pressure, elevated serum cholesterol) and behavioral responses (heavy smoking, drinking, stuttering, dispensary visits).

According to several studies referred to by Van Harrison, the person–environment fit concept has been shown to account for additional variance in strain above that explained by the linear relationships with environment and person separately.

Consistent with Van Harrison's original approach, French, Caplan, and Van Harrison (1982) studied job–person fit and its correlates among more than 2,000 men from 23 occupations in 67 sites. Fit was conceptualized according to the bipartite definition just described. The investigators found that job–person fit was a more powerful determinant of mental health than occupation. Thus, factors influencing mental health varied *within occupations* according to job–person fit more than *between occupations*. This can be seen as reinforcement of our argument that variations in the way work is organized, holding occupation constant, are of prime significance in determining mental health.

The key point here is that within certain bounds determined largely by the nature of the work to be done, there is a discretion involved in the organization of work. Exercised in one direction, this discretion can contribute to mental health among the majority of workers: exercised in another, it can contribute to distress and ultimately to mental illness. However, another perspective on job–person fit is possible. It could be argued and sometimes is, that the obligation to "fit" is upon the individual, not the employer. To some extent, of course, this must be the case. People have an ideal concept of the kind of work they want to do and try to find the environment that will suit them best. However, job-seeking behavior is not entirely nor perhaps even mainly discretionary. That is, job applicants are constrained in their choices by a wide variety of factors, many of which are outside their control. For example, social class, economic status, ethnicity, color, and education all play roles in the kinds of jobs for which people may qualify. Employers located in certain areas may be similarly limited in the kinds of people they may draw upon as employees. These considerations conspire toward a situation in which the employer will often have to make accommodations to the characteristics of the available work force. Consequently, job–person fit would seem to be best perceived as a joint responsibility of employer and employee.

Legal Implications

An increasingly large body of workers' compensation law suggests that employers are now more likely than ever to be seen as having major responsibilities under the law for the mental health of employees.

The evolution of workers' compensation benefits in the United States presents a striking example of the awareness of the role of work itself in mental problems. Originally, only physical disabilities due to organic injury on the job were covered. Gradually coverage grew to encompass psychiatric disorders caused or aggravated by organic injury on the job, physical disabilities originating from job-related emotional stress, and finally psychiatric disorders emanating from work-associated emotional strain (Lesser 1967; Trice and Belasco 1966; Lesser and Kiev 1970; Barth and Hunt 1980). Compensation may be allowable even if the psychiatric condition was latent before and the work situation simply aggravated it. In many jurisdictions employers have to disprove causal connection between employment and disability in mental illness cases, rather than the employee having to prove that such causal connections exist.

In Canada, the evolution of workers' compensation law to cover mental problems has been much slower and shows few signs of speeding up (Ison 1983). Nonetheless, a ground swell of antagonism is developing, evidenced in books such as "Assault on the Worker" (Reasons, Ross, and Paterson 1981) where it is seriously and convincingly argued that violations of occupational health and safety laws should be "criminalized" so that employer negligence would under certain circumstances be treated as crime committed in the pursuit of organizational goals. Accordingly, emotional damage resulting from reckless or negligent behavior on the part of employers and their agents might be the subject of criminal proceedings, particularly where such damage involves major trauma, debilitation, and hospitalization.

Impact of Workplace Innovations on Mental Health

Even the most casual observer can hardly fail to recognize that the workplace is in a ferment of change. While some of this change is unpatterned, a good deal of it is guided by planned efforts to modify the organization of work. The goals of such efforts tend to relate either to improvements in productivity or to improvements in job satisfaction (however defined) or both. Little agreement can be found in the literature concerning the likelihood of achieving both ends through the same means. Katzell and associates (1975) took the view, for example, in their review of workplace innovations that productivity gains were more likely to emerge than satisfaction gains even when both were targeted. In contrast, the studies conducted by the Work in America Institute (book 2, 1978) suggested that work redesign projects were most likely to result in improved satisfaction, with improved productivity as a by-product. It is not fruitful to examine this continuing debate in any depth. We note in passing, however, that the complexity of the variables involved in even the simplest change to the organization of work is so great that only the most heroic efforts on the part of multidisciplinary research teams are likely to even come close to determining cause-effect relationships. On the other hand, EAP and HPP practitioners usually have an intimate knowledge of the workplaces in which they ply their trade so that they are in a position to monitor the impact of innovations upon the mental and physical health of workers in ways that are typically unavailable to external researchers. In monitoring such effects, and in being involved in the planning of innovations, practitioners might consider the use of a criterion or rule of thumb according to which the probable impact of any given change might be predicted. This criterion emerges from our prior discussion of job–person fit: Changes in the organization of work that contribute toward greater consistency between corporate goals, the social and technical systems employed to achieve those goals, and the needs of workers hired to carry them out are likely to contribute to greater mental health among such workers. Innovations that create greater inconsistency can be predicted to have a correspondingly negative impact upon mental health.

This concept of consistency takes stock of the fact that the organization of work occurs in the context of environmental exigencies such as market conditions, economic fluctuations, governmental regulations, and so forth. Furthermore, the available technology for getting work done imposes limits on the structure and process of work organizations. Consequently, the organization of work must be considered benign at least in intent if decisions pertinent to it have been made with the best mental health of workers in mind given the kinds of environmental and technological constraints just mentioned. Nevertheless, the main point is that even within these restraints, employers have some degree of choice with regard to how work is organized, and the direction of this choice has major implications for mental health.

A considerable number of workplace innovations can be identified, all of them testable against the rule of thumb just outlined. The classification of such innovations presents some difficulty, however, since not all of them are logically distinct. White's typology is among the more useful methods of organizing the interventions that have been attempted (White 1979). He indicates two broad classes of planned change—intrinsic and extrinsic innovations. The former refer, for example, to changes in actual work processes such as job rotation, enlargement, enrichment; the introduction of semi-autonomous work groups, quality circles, and so forth. Extrinsic innovations refer to situations in which conditions of work such as the compensation package and hours of work are modified as in the case of flextime, compressed work weeks, and so on. In their review of the probable impact of these and other innovations, Suurvali and Shain (1981) concluded that all of them were of equivocal value with regard to mental health. If a change is instituted in a vacuum, for example, without consideration of how it may affect other aspects of organizational functioning, it is likely that this will insinuate further elements of inconsistency into the goals/means/needs chain of which we wrote earlier. The function of inconsistency is to introduce further stressors into the workplace with which workers must then cope. For example, job enlargement can be seen as compounding the triviality of certain kinds of work and thereby defeating its own objective, which is to make jobs more meaningful. Job rotation, if imposed, may add to the workers' sense of alienation from work. Indeed, rotation appears to be aimed at an assumed need among workers for variety in their jobs. It may, therefore, miss the more basic need for control over the means of production. In so far as any of these techniques can backfire in terms of worker needs, one might expect a greater incidence of mental health problems. Even job enrichment could be associated with such an effect if appropriate, consistent organizational supports are not introduced at the same time. For example, training must be provided in most cases to allow for competent extensions of job skills both in technical and social aspects. Without such support, enrichment could turn out to be highly stressful for individuals operating out of their depth. Similarly, it is not clearly known at what level extensions of responsibilities have to be compensated in order to protect the satisfactions presumably implicit in enriched jobs; that is, a deficiency in extrinsic conditions could nullify the effects of otherwise benign intrinsic changes.

It seems reasonable to predict that the more fundamental an intervention is, the greater will be its potential for producing either strongly positive or strongly negative effects. The actual results might depend not only on the intention behind the innovation but also on the particular way in which it is introduced. Accordingly, we may portray autonomous work groups, and their close relatives, as having the wherewithal to significantly increase the extent to which worker needs are either accommodated or frustrated. At their best, such innovations appear to have the benign effects of job enrichment, and for much the same reasons. However, since by definition the element of group dynamics is introduced and exploited, some riskier aspects become evident. The governing factor behind this riskiness is essentially the content of the normative beliefs, attitudes, and behaviors in specific work group situations. We may see examples of cases where a highly competitive ethic can develop in a group; where conformity becomes highly valued and deviance is punished in subtle or obvious ways. In short, it may not be adequate to create autonomous work groups in an organizational vacuum. It will be necessary to develop strong organizational norms for the social functioning of work groups. These norms will need to be translated into expectations and standards, which would become the yardsticks against which such functioning is in part assessed. It may be recognized that these propositions are closely related to the construct of consistency between organizational goals, methods, and worker needs. If work groups are introduced in such a way that they serve the goals of the organization and the needs of the workers, a state of benign consistency could be said to exist. We hypothesize that this state of affairs will be associated with a lower incidence of mental health problems, according to the theoretical constructs outlined earlier in this chapter.

A particular concern with autonomous work groups presents itself in connection with the identification and referral of problem employees under the auspices of existing assistance programs. The programs usually allow for individuals who are experiencing job-related difficulties, or some problem that might affect their jobs, to seek help under the provisions of a policy or according to procedures that are generally known (Shain and Groeneveld

1980). This voluntary seeking of help may not be seriously affected by the introduction of autonomous work groups. Indeed, it might be facilitated. However, EAPs also provide for more formal referrals of problem employees. This takes place when supervisors—or occasionally the medical department—become aware of serious problems, such as excessive drinking, that are not amenable to normal supervisory intervention but that are clearly affecting job performance in adverse ways. In more formal referral situations, supervisors usually document deteriorating job performance, give verbal and then written warnings. At some point, usually after the second or third ineffectual warning, the employee is given the choice of either accepting a referral to a helping agency or taking the further consequences of disciplinary action. At present, a large unknown must be marked against the question of how such employees are managed in situations where autonomous work groups have been introduced. The issue arises because many supervisory functions are given over to work groups. The supervisor in such situations is often left with "boundary maintenance" functions that involve adequate resourcing of the work groups, coordination of activities, interdepartmental communications, and general planning and accountability. It is not at all clear how existing EAPs apply to these situations. Since work groups have sometimes been given the power to hire and fire—at least in nonunionized settings—one might expect that they also have authority in the area of employee assistance. If this is so, the present writers feel that the negative potential of group dynamics could manifest itself in particularly mischievous ways. We urge, therefore, that such innovations be investigated in future with specific reference to the management of problem employees and employees with problems.

The Socio-Technical Systems Approach

The most basic and far-reaching types of planned change are often referred to as socio-technical innovations (popularly known as "socio-tech"). The term socio-technical implies overhaul of the most fundamental aspects of the work environment, from how the work is physically accomplished to the management of the people whose needs and characteristics must be meshed with the goals of the organization. Historically, there has been a tendency to emphasize the technical system and to take the social system somewhat for granted. Organizations relied on extrinsic rewards to achieve worker compliance with the demands and necessities of the technical system. However, as people become less willing to enslave themselves to machines, they are demanding that organizations give more consideration to their social systems and to the management of human resources (White 1979). As it is impossible to design a plant without simultaneously designing some kind of social organization, whether or not the designers have any explicit intentions in that direction (Davis 1979), it makes sense to consciously try to create a social system that will make effective use of the technical system and provide reasonable satisfaction for its members. The result is the main objective of the socio-technical systems approach: to jointly optimize the two systems (technical and social) so that they are integrated and mutually supportive of each other (Hackman and Lee 1979), thus leading to a more efficient, adaptable organization and more fulfilled employees (Pasmore and Sherwood 1978).

Cherns (1978; see also Hackman and Lee 1979) has outlined nine principles to guide the application of socio-technical ideas to the design of work. These are of some assistance in understanding the specifics of the approach.

1. The process of design has to be compatible with its objectives. For example, if the organization is to be effectively democratic, people must be permitted to participate in designing their own jobs.

2. Only those things that are essential about tasks, roles, and methods should be specified.

3. Variances or unprogrammed events in the work that cannot be eliminated must be controlled as close to their point of origin as possible. For example, production workers should do their own inspection.

4. Jobs should involve many functions rather than highly specialized, fragmented tasks.

5. Departmental boundaries should be located so as to maximize functional autonomy in relation to the tasks being carried out. The boundary maintenance role (for example, personnel and control systems) should reinforce the behaviors the organizational structure is trying to elicit.

6. Information should first flow to the point where action based on it must be taken. Frequently this will be the work group.

7. Systems of social support (for example, personnel and control systems) should reinforce the behaviors the organizational structure is trying to elicit.

8. High-quality work experiences (with responsibility, variety, etc.) should be provided for those who want them while those who do not should be matched with appropriate jobs.

9. The design must undergo constant evaluation and review to determine the need for further redesign.

Those familiar with the popular book "In Search of Excellence" (Peters and Waterman 1982) will recognize the similarity between the principles of management that according to the authors characterize the most successful corporations, and the principles of socio-technical innovation outlined by Cherns (1978), among others.

Many studies have been reported that appear to support the value of socio-technical interventions (for reviews see Katzell, Bienstock, and Faerstein 1977; Hinrichs 1978; Hackman and Lee 1979). The criteria for success usually involve improvements in both productivity and job satisfaction, but the design of the studies, as reviewed by the investigators just listed, often suffers from an absence of comparison or control groups. This lack probably reflects the enormous difficulty of carrying out well-controlled studies of such comprehensive attempts at change. It is almost certain, however, that socio-technical innovations fail miserably in some instances, creating more hardship than originally existed. These cases, not surprisingly but unfortunately, tend not to be reported in the literature.

In some senses, too, socio-technical innovations are likely to be political battlegrounds, particularly when conducted in unionized settings, since such changes may be perceived as attempts to co-opt the union. Indeed, the potential for paternalism in these interventions is enormous when any of the principles outlined by Cherns are ignored (Suurvali and Shain 1981). This should not be read to imply, though, that socio-technical innovations cannot be successfully carried out in unionized settings as Guest (1979), Hinrichs (1978), Glaser (1974), and others have described. However, it would seem highly counterproductive for management to unilaterally introduce major changes that were "for the workers' own good" without careful consultation with employees and their representatives. It is extremely important, for instance, to avoid assumptions about the "meaninglessness" of routine production work to which "socio-tech" is the antidote. Although many of the studies referred to here have supported the view that *any* job under adverse organizational conditions can become meaningless and even degrading, this should not be assumed to characterize any particular job or occupation by definition. The principal criterion for testing the meaningfulness of a job or occupation, as suggested earlier, should be the degree to which it meets intrinsic and extrinsic worker needs as defined by specific job incumbents.

How Can EAP and HPP Professionals Respond?

We began this chapter with the statement that the impact of the work environment upon mental health has major, if perhaps unwelcome implications for the work of EAP and HPP professionals. In the course of reviewing the various bodies of literature that exist on this topic we have suggested certain roles for such professionals that will hopefully allow them to make use of insights gained in the conduct of their everyday work for purposes of influencing the corporate environment and in particular the organization of work. This, of course, is a highly sensitive area for EAP and HPP practitioners, since most are worried about maintaining credibility in their nascent professions and feel they may compromise it if they presume to tell managers how to do their jobs. Indeed, if such practitioners do present themselves as management consultants when they are not qualified in this area, they are certainly gambling with their credibility and ultimately with their jobs. However, such is not our proposal. We suggest rather that EAP and HPP professionals are in a good position to perform the following roles for which they should seek legitimation, some of which are already being played to one degree or another by certain members of the professions in question (see, for example, Francek 1985; and Francek, Klarreich, and Moore, 1985).

1. *Monitoring of environmental impact.* The monitoring role is one in which EAP and HPP practitioners take note of problems that appear to originate primarily in the organization of work and that are reported by or observed in clients seen in clinical, educational, or training situations. At this stage the role is fairly passive in that impressions and observations are merely recorded. A somewhat more active component is introduced when patterns are sought and found, for example, when EAP referrals are emerging from specific departments and presenting with the same or similar work-related problems or when voluntary participants for stress management courses come predominantly from some units rather than from others. This suggests the need for

2. *Feedback to management and union.* As its name suggests, this role component involves the finding of opportunities to inform senior managers and union personnel about the patterns of distress that have been observed and that appear to originate in the organization of work. Ideally, this feedback should be expected and even required by management rather than imposed upon it. Managers should recognize the value of observations about the organization of work emanating from EAP and HPP sources.

3. *Consultation on organizational change.* When changes are being contemplated that affect the organization of work, it makes sense for EAP and HPP professionals to consult on their design and implementation since they are in a good position to predict the probable impact of any innovations upon the mental (and physical) health of workers. Ultimately, such practitioners deal with the fallout from poorly planned and executed changes in the organization of work, so it is in their best interests to have input to decisions at the earliest possible stage. This role is likely to take a long time to become legitimate in the eyes of some senior managers and CEOs. However, such managers will be the poorer for failing to recognize the contribution of their EAP and HPP staff to decisions that, like it or not, affect the mental health of workers.

4. *Design of programs with environmental impact.* When interventions are being designed, in both EAP and HPP areas, they should incorporate elements that recognize the probable impact of the work environment upon the success of both treatment- and prevention-oriented efforts. For example, in the EAP, the principle of job–person fit suggests that if a referred employee

is distressed because of his or her unsuitability to the work, or if the work is designed in such a way as to be unsuitable for anyone, one of the alternative interventions should be to transfer the client to a job whose requirements better match the skills and temperament of that individual. In addition, it would be advantageous for the EAP practitioner to have the ear of those who have the power to redesign jobs in cases where there is evidence that the current design generates more distress than necessary.

Similarly, in health promotion it is important for such interventions to incorporate messages and methods that will help participants to develop a greater sense of self-efficacy or mastery. We noted in chapter 6 the strong relationship between self-efficacy, anxiety, and general distress and observed that program-induced reductions in distress were associated with increases in self-efficacy. Thus, from these two dimensions (self-efficacy and distress) we were able to derive a useful measure of mental health. We suggested further that certain types of HPPs might be able to influence self-efficacy directly rather than trying to deal with distress and its symptoms, even though ultimately distress would be reduced given the aforementioned correlation. The means of achieving such ends are still very much in the developmental stage, but they are hypothesized to include the design of educational and training programs that demonstrate to participants that significant health-related changes are achievable through (a) small, self-reinforcing steps toward behavioral objectives in specific areas, (b) recognition of environmental constraints upon choice in health-related decisions, and (c) relatively minor changes in such environmental conditions. The last two steps involve some element of consciousness-raising in the sense that HPP professionals may need to develop an awareness among participants of how their health-related choices are constrained by the organization in which they are employed. We do not intend to suggest by this that practitioners bite the hand that feeds them. Indeed, if management does not value the raising of a critical consciousness among participants in HPPs and practitioners do it anyway, they will probably be out of a job in record time. However, if management does value this role and encourages feedback based upon this critical examination of environmental constraints, the organization will have at its command a powerful vehicle for self-regeneration and reform.

Some such philosophy appears to have guided the design of Johnson and Johnson's "Live for Life" program (Wilbur and Garner 1984). One outcome of this process was reportedly the introduction of flextime to accommodate the needs of employees whose stress levels had become intolerable as a result of trying to juggle timetables in two-career families, including the all-important child-daycare arrangements. The skills involved in the performance of this difficult "catalyst" role in relation to environmental change are as yet hardly defined, yet it is essential in our view that they be developed since they relate to a fundamental ethical question in health-care delivery through the workplace, namely, in whose interests are HPP practitioners acting? If they are content to ignore the kinds of environmental issues that have been raised in this and earlier chapters, it could be argued that they are acting mostly on behalf of the employer and very little on behalf of employees.

5. *Reinforcement of programs through social support.* In relation to the role just outlined, an increasing body of knowledge is being accumulated and disseminated on the value of social support networks in the workplace. These networks may serve to reinforce the coping strategies of individuals as they seek to deal with the various environmental stressors outlined in this chapter. The formation of self-help groups around specific areas of concern and the use of informal networks to bolster efforts to change health-related behavior involve skills that some EAP and HPP practitioners already have but many probably do not. This set of skills is required to perform a role that could contribute significantly to the success of currently available EAPs and HPPs. This potentially important and complex issue is taken up in chapter 9.

9

The Impact of Social Support on Health: The Role of EAPs and HPPs

In the last chapter, it was suggested that a natural liaison exists between the informal social systems that may be found in all organizations and the work of EAP and HPP practitioners. This liaison is natural in the sense that the formal provision of health-related services through the workplace interacts with the informal social system regardless of whether practitioners intend it or not. The social system must be presumed to exist. It is the medium through which any formal workplace intervention must find its way, and, accordingly, it can either facilitate or thwart the efforts of health care providers. Thus, an understanding of the functions of informal social systems with regard to the needs of their members is important for EAP and HPP professionals.

Informal social systems originate from the needs of members for a commodity called "support," the nature of which can take many forms depending to a large extent upon the characteristics of members and the situations to which they must adapt. Thus, depending on the fit between worker characteristics and organizational conditions, the resultant informal social system may emerge either as a structure that generally promotes the achievement of corporate goals or as one that expends the majority of its energy on defense against real or perceived corporate abuses.

Social support serves a variety of functions, not always simultaneously and not always in the same degree. Cobb (1976), for example, considers that social support is generated when an individual receives information from others that leads him or her to believe that he or she is cared for, loved, esteemed, and valued and that he or she belongs to a network of mutual obligation and communication. Thoits (1982), in a similar vein, argues that "support is the degree to which an individual's needs for affection, approval, belonging and security are met by significant others." Turner, Frankel, and Levin (1982) stress that support is provided when it leads to a sense of attachment or belonging, when it contributes to social integration, provides an opportunity to nurture others and to receive guidance, reassures the individual as to his or her worth or competence and gives a sense of "reliable alliance."

Implicit or explicit in these definitions of support is the notion that the benefits of this commodity are at least to some extent earned through the discharge of obligations to, or fulfillment of, expectations held by other members of the network or group. Indeed, in some cases it is clear that the obligations of group membership exceed the benefits, as Belle (1982) and others have pointed out. Too much is put out (or is perceived to be put out) for too little in return. In the tradition of research known as the "social network" approach, the social ties of an individual are seen as defining collectively an economy of exchange in which most interactions have both costs and benefits. In other words, the network approach does not assume that social ties necessarily provide support in the terms described earlier (see Mueller 1980; Lin, Dean, and Ensel 1981; Hammer 1981 for expositions). According to this perspective, social ties are plotted and their potential for support is examined. Networks can be conceptualized in terms of size; frequency of contact between members and the focal individual; proportion of kin to nonkin; geographic distribution; and directionality and reciprocity of the interactions within them. In addition, the *density* of relations is considered to be important. This refers to the degree of association not only between members and the focal individual but also between the members themselves (Hammer 1981, 16). A final dimension concerns the proportion of so-called "*multiplex relations*" within the network, that is, how many relationships cut across more than one content area (work, home, recreation, sport, community, church, etc.).

The costs and benefits of group or network affiliation are matters to which we shall need to return. However, for the moment, let us assume that support can be and is generated in varying degrees by informal and sometimes even formal social systems. What, then, are the relationships between such support and the health of those exposed to it as well as between all this and the work of EAP and HPP professionals? These questions must be addressed sequentially since the answer to the latter depends upon the former.

Social Support, Stress, and Health

The two major hypotheses linking social support and health both involve the relationship of the former to stress. In one conceptualization, social support *mediates* or *buffers* the impact of stressors upon health. It is, therefore, referred to frequently as the "buffering hypothesis" (see Cassel 1975; House and Wells 1978; and Simpson 1980 for details). Thus, support can provide some degree of predictability, purpose, identity, meaning, and hope in situations where work is organized in such a way as to produce the opposite of these states, conditions, or circumstances. In another conceptualization, social support is seen as having a *direct* impact upon health. That is, deficiencies in support are thought to act as stressors in themselves regardless of the presence or absence of other stressors in the environment (Mueller 1980). The "direct effects hypothesis" (just described) and the buffering hypothesis are not exclusive of one another. Caplan (1971), Gore (1978), Barrera (1981), and Frydman (1981) are among a growing number who argue that both hypotheses may be valid, a view that we are inclined to share.

Insofar as support buffers the impact of other stressors, it is believed to operate by (1) creating normative pressure (through family, friends, etc.) against the occurrence of stressful events, for example, divorce, (2) altering perceptions of threat from the environment, (3) encouraging functional coping responses (through neutralization of, adaptation to, denial of, distraction from, defusion or deflation of the stressor), (4) encouraging problem solving (instrumental solutions), (5) helping the individual to deal with the emotional consequences of a stressful event (for example, death of a loved one), and (6) reinforcing tolerance for strain or resistance to it through exercise, nutrition, relaxation, rest, and so on.

Insofar as social support is directly linked to health, it appears that an important dynamic of this relationship evolves from changes in the amount of support that is perceived to be or is actually available. Thus, in this approach, death of those near and dear, divorce or separation, illness, change or loss of employment, marriage, birth, relocation, and other major life events are significant primarily in terms of the changes that they bring about, for good or ill, in the support systems of affected individuals (Mueller 1980). Lowenthal (1964) suggested in effect that such changes represented a disruption of the individual's stable state requiring adaptation to added or reduced levels of social support. In a sense, then, the whole tradition of stress research that sees life events as stressors linked to health (see chapter 6 for more discussion) is hereby placed in the context of social support research. At the very least, it seems clear that an awareness of the impact of changes in social support has been frequently lost in studies of how life events affect health (Schaefer, Coyne, and Lazarus 1981).

There is, of course, a circularity in this approach under certain conditions. Changes in an individual's health status are in themselves life events that, by positively or negatively affecting the availability of social support, further influence health status. This awareness has become the basis of the "reciprocal effects hypothesis" (Henderson 1980; Turner, Frankel, and Levin 1982; McFarlane et al. 1983; Ferraro, Mutran, and Barresi 1984). A particularly complex example of reciprocal effects occurs in the case of mental illness where it is possible that deficiencies in support or in the ability to utilize it contribute to deteriorations in mental health, which tend to further isolate affected people from the nurturance or other help they may require in order to recover. Mentally ill or disturbed people may then self-select into environments from which little or no assistance is to be obtained.

The pathway between social support and health, whether we accept a direct, buffering, or combined model of the relationship, is clearly routed through the domain of stress. However, this begs the question of how the experience of stress itself affects health. In this regard the impact of the bio-chemical products of stress reactions upon the immune system is probably a major factor (see chapters 2 and 6 for further discussion). Whatever the mechanism linking support and health, few now question the reality of the connection; so much so in fact that some investigators define the empirical question as "under what conditions and in what direction does social support affect health" rather than "does the one affect the other at all" (House, LaRocco, and French 1982). This trend toward redefinition of the research task has been accompanied by a growing recognition of the idiosyncratic nature of support requirements by individuals. Thus, a notion of "optimal support requirements" regulates the investigations of those who consider that perceived support (subjective measures) is more important than so called ob-jective measures of the same commodity (Barrera 1981; Schaefer, Coyne, and Lazarus 1981; Ward, Sherman, and LaGory 1984). While variables such as age, sex, marital status, socioeconomic status, ethnicity, cultural background, employment status, and so forth are seen to influence the nature and degree of support required, they do not in themselves define the need. Only a knowl-edge of individual needs will provide a valid and reliable picture of support requirements sufficient for useful action to be taken (Blythe 1983).

Social Support Systems and EAP/HPP Service Delivery

As noted at the beginning of the chapter, EAP and HPP professionals interact with social support systems reactively or proactively in such ways that the effect of their work is either potentiated or minimized in specific circumstan-ces. Gottlieb (1982) describes how people trying to deal with emotional prob-lems interact with the informal social system surrounding them and how this interaction affects the efforts of formal care givers such as EAP and HPP practitioners. Gottlieb sees four stages in this process. The first involves prob-lem recognition and crystallization. During this phase, problems are infor-mally diagnosed, accurately or erroneously, by the subject's network associ-ates. The employee's acceptance or rejection of the diagnosis is influenced by the *credibility* ascribed to members of the informal support system. High credibility can be anticipated when members provide support to the focal individual in the terms outlined earlier, that is, he or she receives affection, approval, a sense of belonging, and security. However, even the provision of these commodities does not influence in any major way the likelihood that messages received from the group will be consistent and clear. Indeed, the advice of one's support group can paralyze rather than motivate a troubled person who is trying to determine what to do about his or her problem. This complicates the second stage, which involves help seeking, the sorting out of how and from whom to get help. Again, the consistency issue arises since advice from the support group on this subject may vary considerably. Even when help is actually obtained (assuming this stage is reached), the troubled person may receive conflicting counsel from formal and informal care givers during this third phase of the coping process. During reintegration and normalization (the fourth stage) this potential conflict may again manifest itself.

Sonnenstuhl (1982) also emphasizes that the social network can help direct people away from as well as toward formal sources of aid such as EAPs provide. Once the troubled individual is in the formal system, the informal network continues to pursue its own agenda whether or not it is consistent with what professional care givers consider to be appropriate. From the formal service provider's point of view, then, it would be advantageous if the informal network (a) contributed to the motivation of troubled employees to seek help from the EAP (or HPP) when their own efforts appear in themselves to be no longer sufficient, and (b) supported the EAP in its approach to problem resolution. This, however, is easier said than done. Owing to the necessary constraints of confidentiality, EAP practitioners cannot share with the client's network the nature of presenting problems or the plan of action agreed upon with the client, except where members of the support system are formally involved in treatment with the client's consent and the service provider's agreement. This does suggest, however, that thought might be given to the involvement of at least mutually agreeable members of the client's network in at least certain phases of treatment. The purpose of this involvement would be to provide further perspectives on the client's problems, to consider creative approaches to their solution, and to raise the chances of consistency between the actions of formal and informal care givers.

Prior to getting people involved in the EAP at all, however, the issue is more one of disseminating accurate information about formally available services to as large a proportion of the work force as possible so as to saturate the informal networks with an understanding of how their efforts can mesh with the EAP or with the HPP in cases where the latter offers skill-building courses such as stress management, assertiveness training, smoking cessation, and so on. Of course, this proposal possesses merit only where the EAP or HPP can justifiably lay claim to the trust of the work force and its constituent networks, many of which may have differing expectations about how care should be provided. The level of trust that can be developed in this regard is not just a function of how sincere or competent the EAP practitioners are: there are many snares and pitfalls along the way between a desire to seek help and making contact with the people who can provide it. These obstacles can take the form of hostile managers and supervisors who do not accept the program or have not been trained in its use; medical and personnel departments who do not understand or share the philosophy of EAP; personnel policies or benefits packages that make access to the EAP difficult or impossible; unions and members who distrust the motivation of management in setting up assistance programs, particularly when they were not consulted. With information at its command about these and other organizational blocks, the informal social system may well rally around a virtual sabotage of the EAP or HPP by diverting its members away from all formal sources of help.

The implications of such a state of affairs for service providers are basically those drawn out in chapter 8, where EAP practitioners were seen as having important roles to play in modification of environmental conditions that they identify as being adverse to mental health. In that sense, an important, indeed crucial part of an EAP is the management context within which it operates. A workplace in which the organization of work is inimical to mental health may either subvert the informal social system altogether or create the kind of reactive, defensive "inmate culture" syndrome. In either case, the EAP or HPP professional will find that an essential medium for his or her work has been severely weakened or soured by workplace conditions.

It is conceivable, however, that even under such adverse conditions, EAP and sometimes HPP practitioners could assist individuals who managed to connect with them by paying attention to the client's social support system. This does not mean attempting to manipulate the system directly, but rather helping the focal individual (client) to evaluate his or her own network to determine its strengths and weaknesses (Blythe 1983). Cobb (1976) suggests that individuals may be trained in the giving and receiving of support, while Steinert (1978) recognizes that some people do not know how to identify sources of support or how to use them. House and Wells (1978) are dubious, however, about the value of carrying out this kind of "support training" if the environment is extremely stressful, on the grounds that we should not be pulling people out of the river when we would be better occupied in stopping them from being pushed in. This objection is all very well, but interventions such as stress management and support training are forms of self-defense that can be implemented while efforts at more basic organizational change are underway. It is possible, too, that an EAP could be in the vanguard of such change at least in situations where management is able to recognize the value of feedback received from helping professionals who report they are spending a good deal of their time patching up unnecessary casualties and attempting to shore up their clients' natural but beleaguered social support systems.

In this regard it is worth recalling that some of the more fundamental work innovations described in chapter 8 under the aegis of socio-technical overhauls are centrally relevant to the reinforcement of social support. Even in these situations, however, dangers were apparent in that such interventions often tamper with existing support networks, attempting to replace them with newly contrived units. Semi-autonomous work groups are examples of an approach that could backfire if introduced without careful attention to the natural support systems of the people whom they affect. Nevertheless, the principle of worker participation that underpins the socio-technical approach is clearly related to the dynamics of social support described earlier in that it tends to improve employees' sense of control or mastery over their own lives. This sense of self-efficacy is, as we have seen, related to anxiety and to other symptoms of distress. Thus, Gottlieb (1981) together with Crouter and Garbarino (1982) consider that a basic primary prevention task in the workplace is the facilitation of social interaction from which social support may flow.

While, as noted earlier, EAP/HPP professionals may be able to have considerable impact at this basic primary prevention level through direct input to organizational decision making, a more grass-roots-oriented approach is sometimes likely to provide a slower but perhaps surer route to the same objective. This has been evidenced in several different interventions, which we refer to briefly in the following.

The "Lifegain" Approach

Allen and Linde (1981) describe a method of introducing health promotion to the workplace that operates on the basis of changing the *health culture* of organizations. The basic building blocks of this approach are the support groups that are formed around topic areas such as nutrition, dieting, exercise, alcohol and tobacco use, stress management, and so on. Prior to the introduction of workshops, however, a considerable amount of preparation is required in terms of top organizational level support. Once this is obtained, with as many visible manifestations of senior management and union approval as possible, a process is begun in which natural leaders and a volunteer network are identified. It appears that during this initial phase a formal or informal survey of the existing health norms, practices, and reported problems of the work force is carried out. The authors of this approach supply questionnaire devices that may facilitate or direct this process. The results of the survey are used to indicate the areas where introductory workshops might be brought in. The content of these and later sessions emphasizes personal empowerment through development of an understanding of how the health culture affects individuals and of how this culture can be modified through collective action. People are also taught that working on one area successfully is likely to have significant ripple effects in other areas. Essential to the follow-through from workshop sessions is the formation of support groups developed around specific health-related concerns. These are self-regulated groups consisting of between eight and twelve members. They vary in function depending on member needs and the subject area, from providing mainly expressive support to supplying mainly instrumental support. It appears that these support groups are essentially seen as cells that proliferate throughout the organization. Participants are encouraged to involve themselves in outreach activities, forming larger groups or task forces for specific purposes such as the acquisition of resources (for example, facilities, child daycare, professional help) or to put pressure on management to change environmental factors such as smoking rules, cafeteria food, and more basic conditions such as hours of work.

While the "Lifegain" approach possesses conceptual merit in terms of developing social support, we would argue that sound evaluations of the method are needed. Many informal reports of its success exist, but little attention has been paid to possible negative outcomes or indeed to objective measurement of positive outcomes. With regard to the negative potential, we observed earlier that the overlay of new "support" systems upon an existing informal support network could disrupt the old without effectively introducing an alternative. In addition, certain insidious possibilities present themselves. One is the pressure, real or imagined, to become involved in what sometimes sounds like an evangelical movement. Another is the threat that may be experienced by those whose existing social relations, even intimate relations, may be strained by conflict between new and old health norms, practices, beliefs. Yet another danger is the isolation of those who perceive their social skills to be inadequate for participation in the cell groups. Some of these dangers are similar to those we outlined in relation to socio-technical innovations here and in chapter 8.

Johnson and Johnson's "Live for Life" Program

An account of this program is given by Wilbur and Garner (1984) in the context of social marketing theory. This approach requires that employee needs be considered as the basis for health-related programing. In this sense, the Johnson and Johnson program superficially resembles the "Lifegain" method. However, the presenting problems of the host company when setting out to consider health promotion were escalating health-care costs that were believed to be related to deficits in the life styles of employees. Thus, there may be some question whether the employer entertained any doubt that a program of some kind would go ahead.

Indeed, the initial vision of the project required that changes in life style would have to be supported by changes in the environment, which meant that the maximum number of employees should be involved in the process. This, again, is a reprise of the "Lifegain" approach but with a difference—the difference lying in the obviously superimposed notion of changing the health norms of the organization through managerial fiat. Readers may recall from chapter 1 how historically a seemingly benign idea such as health promotion can be colored in threatening hues when refracted through the ideologies of individual paternalists. In this case we have a benign idea, the reinforcement of social support, converted into a not-so-benign idea, the exploitation of social support. Within this somewhat ominous framework, a number of otherwise conventional approaches were used toward the promotion of health. A "health screen" was run on 76 percent of all 35,000 employees in seventy-five locations to gain some measure of health status. This was followed by a life-style seminar, which was an introductory overview of the whole program. The screen (with feedback on results) and the seminar acted as motivators and funnels into the focused HPPs in nutrition, weight control, smoking cessation, stress management, and exercise. In these programs, there was an emphasis upon feeling better immediately and upon having more energy, rather than upon disease prevention and longer life even though the latter were corporate objectives.

A vice-president of the company was commissioned to head the program and to develop a volunteer network to implement it throughout the seventy-five locations. This person was assisted by a small core of paid people at corporate headquarters. One of the stated reasons for having a vice president head the program was that he or she would provide a direct conduit to the highest level of management so that feedback from the grass-roots level on matters pertaining to the organization of work could travel unimpeded. It is reported that flextime was introduced as a result of such feedback, employees having observed—presumably in a rather forceful way—that a considerable amount of the stress perceived in relation to home–work conflict could be resolved through these means. According to preliminary reports, 65 percent of the work force was reached by the life-style seminars (more specifically, 23 of the 25 percent targeted overweight employees, 17 of the 20 percent targeted for stress management, and 20 of the 20 percent targeted for smoking cessation). It is not reported whether these targets were based on knowledge of the identities of people known to be at risk according to the health screen or whether a simple statistical association was made. No firm data were available at the time of this report to indicate whether there were lasting results from the program at an individual level or whether the objective of fundamentally restructuring the social environment at work was reached (Wilbur and Garner 1984, 153–54).

The Wallingford Wellness Project

Blythe (1983), who recommends training in the giving and receiving of social support, describes an intervention in which participants in health-related workshops are asked to bring with them some members of their support network. The three-hour meetings focus on nutrition, fitness, stress management, assertiveness training, and related topics. The purpose of involving the support group is to create a climate of acceptance for changes that the focal individual might want to make and to help significant others understand the kind of assistance that will be required from them if these changes are to be maintained. In addition, the workshops are designed to foster environmental awareness so that participants are inspired to advocate for healthier conditions in the community and the workplace. These groups are meant to serve a variety of expressive and instrumental functions. As such, they may well be prototypes for the kind of program referred to in chapters 2 and 5 where the need for "critical consciousness" in health promotion was emphasized. Blythe's method is one that can probably be incorporated into either the "Lifegain" or the "Live for Life" approaches sketched earlier.

Self-Help Groups

Self-help groups seem to develop around several major areas, although their orientation to the subjects in hand varies enormously, sometimes providing room for professional assistance and sometimes not. The compulsion/addiction class of self-help groups is probably one of the largest, at least in Canada where Romeder (1982) conducted a major survey of them. Included in this category are, of course, Alcoholics Anonymous, Al-Anon, and Alateen, various local and networked smoking cessation groups, as well as diverse weight-reduction and eating disorder groups. Another major class of self-help group involves people united by specific health problems such as heart disease, epilepsy, diabetes, kidney disorders, cancer, and so forth. Problems associated with parenting draw many people together to learn skills, form daycare support networks, brainstorm creative solutions to home–work conflicts, and generally share the load. In addition, a wide range of such groups exist around women's issues, social action, stress management, and mental illness.

Not all of these groups are likely to be specific to single organizations: some, indeed, are more likely to be community-based. The reasons for this are many, but principally concern the prevalence and sensitivity of the problems or issues to which the groups respond, that is, it is unlikely that a group advocating the rights of homosexuals would be company-specific unless confidentiality (assuming it was wanted) could be assured. The fact that some such groups' membership might not be restricted to one organization is not, however, a rationale for noninvolvement by EAP/HPP professionals. After all, practitioners in these fields as a matter of course refer employees to community agencies, so why not to cross-organizational self-help groups? Some students of self-help conclude that even though such groups form often as a result of *frustration* with professional services, they can still profit from professional involvement under certain conditions and circumstances (Froland et. al 1981, for instance).

If EAP and HPP professionals involve themselves in the development of self-help groups, it has been suggested that they cleave to sound community-development principles. Thus, Bakker and Karel (1983) recommend that professionals should decide upon their role in consultation with the incipient group, should avoid leading the group, and should plan to phase themselves out or into different roles as soon as agreed upon mileposts have been reached. Wollert and Barron (1983) see professionals helping in the crystallization of groups, mediating with outsiders on their behalf, acting as brokers with the professional community, providing information and resources, developing skills, and doing supportive research.

Gottlieb (1982) found that many groups can use professionals effectively. The problem may be more with professionals who are sometimes uncomfortable working with self-help groups (for example, Lavoie 1983). Indeed, problems are to be expected. The norms of professionals from various disciplines can be easily at odds with those of self-help groups leading to conflicts or inconsistencies that create anxiety and frustration among members as well as to ineffectiveness. To the extent that such groups serve an expressive support function, professionals may relate to them in an overly instrumental, task-oriented fashion, thus subverting their most valuable role. On the other hand, the introduction of an instrumental dimension at the right moment can revitalize a group that would otherwise founder.

Parent Support Groups

Although we have mentioned them previously, parent support groups will repay our somewhat closer attention because they illustrate and focus a number of the issues raised in this chapter. Such groups may be formed, as already observed, around a variety of parenting concerns. Communication skills, information exchange, resolution of conflicts between work and home, daycare networks, and so forth, constitute a majority of these concerns and provide opportunities for EAP/HPP practitioners to assist employees in a number of ways through group facilitation, research, brokerage, representation to management, lobbying, and so on. Employers appear to be increasingly responsive to arguments that urge them toward an active role in supporting their employees in their capacities as parents, for example, a survey by the National Employer Supported Child Care Project in Pasadena, California showed a 400 percent increase in company offered child care programs between 1978 and 1982 (Fenn 1985). The same author describes the experience of Nyloncraft, a plastics injection-moulding company with an on-site daycare facility. She reports that turnover was substantially reduced among the 450 employees, 85 percent of whom were women, after this service was introduced. Retraining costs were correspondingly cut and greater predictability restored to production. The 24-hour center was opened up to the community, whose children make up two-thirds of the client population.

It is clear that not all, nor even most, companies will be able to offer on-site daycare, but there are a number of other things that employers can do to relieve the stresses that devolve upon their workers as a result of child-rearing or child-care worries. As Coolsen (1983) suggests, flextime, permanent part-time, job-sharing, and work-at-home options are all approaches worth considering if employers want to reduce lateness, absenteeism, poor productivity, and turnover resulting from problems with child care. We have already mentioned the use of parenting seminars and workshops. These may give rise to self-help groups or task forces that can lobby employers (with or without EAP/HPP staff help) for the more structural changes just listed. It would seem to make sense for helping professionals in the workplace to be involved in child-care issues, since these concerns will often be behind other problems. Anxiety, depression, fatigue, and a multitude of other distress symptoms are likely to mask the common reality of working parents' energies being spread too thinly by the conflicting demands of work and home. Given the fallout from these conflicts, employers are probably unwise to segregate home and work issues, yet their involvement in parenting takes us one step closer toward the corporation as a kind of "prosthetic family," to borrow a phrase from Glossop (1980). The Nyloncraft example cited earlier is in this regard a timely reminder of the fact that paternalism is still alive and well in the last part of the twentieth century. At Nyloncraft, on-site daycare operates in the context of a company that offers Mother's Day presents to all eligible women, gives corsages to all female employees, and holds a three-day annual picnic in which helicopter and elephant rides are available, together with beer on tap. Here, the personnel manager operates as a sort of den mother to whom all employees are invited to confide their worries. This is not an EAP.

Conclusion

We have observed in this chapter that social support is essential to physical and mental health, although we must be careful to add that support seems to be mainly in the eye of the beholder. We suggested that social support has a direct impact upon health and that it also buffers or mediates the impact of environmental stressors. In regard to both pathways of action, we urged EAP and HPP professionals to consider the reinforcement of social support as a role that they might fruitfully play, while continuing to pay attention to the neutralization of unnecessary stressors implicit in the organization of work. Various methods of introducing support dynamics into existing or proposed programmatic interventions under the auspices of EAP and HPP were mooted. In addition, the use of grass-roots approaches was discussed, the purpose of which (grandiose though it may seem) is to modify the health culture of organizations. Thus, a number of levels and points of entry are evident when confronting the challenge of improving social support in the workplace. The significance of meeting this challenge for EAP/HPP practitioners is that social support represents a bridge between the *programmatic* interventions typically carried out by such professionals and the *structural* problems that they frequently perceive to underlie many of the apparently individual, personal, familial, or work-related problems that are presented to them in everyday practice. The reinforcement of social support represents a strengthening of this bridge and thereby of the likelihood that the work of EAP and HPP practitioners will be magnified in its effects.

Appendix
Take Charge: A Self-Help Course in
Feeling Better and Living Longer

Outline of Sessions

Session 1. Cardiovascular Health and Its Maintenance
(70 minutes)

By virtue of its introductory nature there were some unique events in this session. These were:

1. Opening statements and endorsements of the course by management and union representatives.

2. Administration of a pre-course version of the questionnaires upon which the data reported in chapter 5 are based.

3. A brief outline of the course was presented by the course convener. During this outline the main themes of the course were established. They may be summarized by the following "message statements," which are reiterated throughout the session both verbally and in written form on flip-charts, wall-posters, and in handouts.

> Your health is your most important resource. You are in control of your health in some very important ways.
>
> Every aspect of our lives is connected in some way to every other aspect—our eating habits, drinking, smoking, obtaining exercise, our way of working, obtaining rest, and recreation all depend on one another to some extent. Small, consistently implemented changes can have far-reaching effects, for example, drinking less coffee, alcohol; eating at regular hours; not eating dessert every day; not adding salt to a meal already cooked; walking to the store instead of driving; climbing stairs instead of taking the elevator; eating fruit instead of candy during office hours, and so on.
>
> Control over health and feeling well can be gained or regained in small steps.

Many of us value "moderation in all things" but it is not always clear what moderation means.

How much drinking is moderate?

How much exercise is enough?

What is a sensible (moderate) amount of stress?

Remember: most of us do at least a *few* things that are "good" for us. We can tip the balance toward doing more things that are good for us by taking small, but consistently implemented steps. In that way, we can protect and take charge of our health.

Following this outline, the convener explained that before each session there would be brief "*focusing exercises*" that would help orient participants to the forthcoming material and help establish the personal relevance of the information presented. These focusing exercises took a different form for each session but they were all loosely based on a self-confrontation technique adapted from Rokeach (1979).

At the end of each pair of sessions it was originally intended that an adaptation of Janis and Mann's "Decision Balance Sheet Procedure" (1977) would be used as a "*synthesizing exercise*." The purpose of such an exercise was to tie in the content of the previous two sessions to the lives of individuals by asking them to identify specific areas in which they would like to make changes and make proposals for how they would go about making them. The structure of this technique is shown at the end of the description of sessions because we believe it is important. However, it was not consistently implemented because of time constraints.

Material from the focusing and synthesizing exercises was reproduced on flip-charts and posters.

Focusing Exercise, Session 1. Convener asks participants to do a private mental exercise according to these directions:

Ask yourself: Am I satisfied with: my present body weight? The amount of physical exercise that I do? Is my diet as healthy for me (and/or my family) as I would like it to be? Do I take the time to relax often enough?

Film, Session 1: "Healthy Lifestyles," American Medical Association. *Synopsis:* Having defined "health" not simply as a lack of disease, but as long-term, overall mental and physical well-being, the film suggests basic changes in life-style patterns that are instrumental in achieving this state. A consideration of the most common diseases clearly indicates that many can be prevented by avoiding known risk factors: lack of exercise, stress, and habits on which we become dependent such as smoking, drinking, and overeating. In a detailed examination of each of these, the processes whereby they can be detrimental to health are illustrated, and constructive ways in which they can be modified are suggested. Interviews with a number of people who have made a successful transition to a healthy life style serve to indicate that by taking responsibility for their own health, people can achieve a new state of well-being.

Session 2. Cardiovascular Health (Speaker) 45 minutes

Focusing Exercise. Convener asks participants to consider privately,
   "On a scale of 1 to 10, how much control do you think you have over the health of your own heart?"

Presenter: A Local Physician Who Is Head of a Sportsmedicine Clinic (Introduced by Convener). *Synopsis:* The presenter, using an easel, lists the major and minor reasons for poor cardiovascular health. The major reasons are identified as family history, smoking, hypertension, and high cholesterol, and minor reasons are identified as stress, gross obesity, and lack of exercise. With the exception of family history, all of the above-mentioned factors are modifiable. After offering numerous alternatives and recommendations the presenter responds to questions from the audience.

Synthesizing Exercises.

Session 3. Alcohol Abuse (35 minutes)

Focusing Exercise Convener asks participants to do a self-inventory based on these directions.
   Ask yourself: Do I usually drink to help cheer myself up? Do I often drink when I am tense and nervous? Do I sometimes drink to help me forget my worries? Do I drink to change the way I feel?
   If you answered yes (or even "maybe") to any of these questions, or if you know someone who would, the film will be of some interest to you. (The items listed are from Parker and Brody [1982].)

Film: Alcohol Abuse; Early Warning Signs (23 min.). *Synopsis:* Hosted by Henry Fonda, the film lists ten of the early warning signs of alcoholism. The ten signs are dramatically illustrated: increased tolerance, increased desire for alcohol, lack of control over drinking, personality changes, drinking alone, increased dependence on alcohol, neglect of family, loss of pride in job, inability to quit, and physical and emotional deterioration. Fonda urges anyone having any of these signs or knowing someone with these signs to seek help immediately. Some sources of help are also given.

Session 4. Alcohol Abuse (Speaker) 45 minutes

Focusing Exercise. Convener distributes a device called "Know the Score." It is a self-inventory of drinking behavior designed and produced by the Addiction Research Foundation. The tally indicates drinking "risk level." Participants do not declare their results to the group.

The convener emphasizes the following aspects of moderate or low-risk drinking.

1. Two standard drinks per day = 1½ oz. spirits × 2 OR 5 oz. unfortified wine × 2 OR 12 oz. regular beer × 2 OR various combinations of any two.
2. No more than 10 percent of daily calorie intake.
3. One standard drink per hour to avoid drinking/driving law violation. Wait one hour after the last drink.

Presenter, Harry Hodgson (Addiction Research Foundation). *Synopsis:* The speaker relates back to the participants' scores on "Know the Score" that he places in the context of the probable physical effects of alcohol on the body over time. He emphasizes the role of alcohol in many of our everyday social and work-related functions.

Synthesizing Exercises.

Session 5. Stress 35 minutes

Focusing Exercises. Convener directs participants' attention to the following questions:

> *Do any of these statements apply to you?*
>
> I am more likely to have a drink than to do something active when I feel tense.
>
> It's hard to turn my thoughts off long enough to relax.
>
> I hardly know when stress is getting a hold on me (until it has happened).
>
> I need sleeping pills sometimes.

Affirmative answers to one or more of these questions are suggested as good reasons to take the upcoming film seriously.

Film: "The Stress Mess" (Barr Films, Pasadena, Cal.), 24 mins. *Synopsis:* Through humorous illustration this film suggests how to identify the sources of stress in our lives and shows many of the common signs of stress. Several major remedial approaches are shown, among them being: setting priorities, delegating tasks, self-affirmation (learning to say no), and learning how to relax.

Session 6. Stress 45 minutes

Focusing Exercise. Convener asks participants to ask themselves:

1. Where do you feel most of the stress in your life comes from?
2. How would you rate your ability to handle stress on a scale of 1 to 10?

Presenter, representative of Sportsmedicine Clinic, Windsor, Ontario. *Synopsis:* The speaker employs a simple "Life Events" inventory that he shows on an overhead projector to illustrate the myriad sources of stress in our lives. He distinguishes functional and dysfunctional stress, illustrating the consequences of the latter with examples of stress-related diseases and maladaptive behaviors, including excessive drinking, absenteeism, and violence. He suggests alternative ways of responding to and managing stress, with an emphasis on the benefits of regular exercise.

Synthesizing Exercises.

Synthesizing Exercise Material: Used after Each Pair of Sessions

In relation to the subject you have just heard about, think of some area within it where you would like to make a change—any change, no matter how small.

Area

Example: *Getting enough sleep* is an area in which some people want to make a change.

Think of an action that could lead to your getting more or better sleep.

Method

Example: Go to bed earlier; get up later, unplug the phone or take it off the hook; have spouse take care of the kids every other morning; refuse extra work.

Now write down the change you would like to make on the paper provided.

Underneath, on the lefthand side of the page, write a heading, "Advantages"; on the righthand side, write another heading, "Disadvantages."

If the disadvantages outweigh the advantages, think of another action that might achieve much the same objective, and count up the pros and cons again.

References

Abrams, D. B., and G. T. Wilson. 1979. "Effects of Alcohol on Social Anxiety in Women: Cognitive Versus Physiological Processes." *Journal of Abnormal Psychology* 88(2):161–73.

Addiction Research Foundation. n.d. "An Initial Interview for Clients with Alcohol-and/or Drug-Related Problems (The 'IICAP')." Addiction Research Foundation, Toronto, Canada.

Alderman, M., L. W. Greene, and B. S. Flynn. 1982. "Hypertension Control Programs in Occupational Settings." In *Managing Health Promotion in the Workplace,* edited by R. S. Parkinson et al., 167–72. Palo Alto: Mayfield Publishing.

Allen, H. 1949. *The House of Goodyear.* Cleveland: Corday and Gross.

Allen, R. F., and S. Linde. 1981. *Lifegain: The Exciting New Program That Will Change Your Health—and Your Life.* Morristown, N.J.: Human Resources Institute.

Allison, K. 1982. "Health Education: Self-Responsibility vs. Victim-Blaming." *Health Education* (Spring): 11–13.

American Psychiatric Association. 1980. *Diagnostic and Statistical Manual of Mental Disorders* (3rd ed.). Washington, D.C.

Armstrong, P., and H. Armstrong. 1978. *The Double Ghetto: Canadian Women and Their Segregated Work.* Toronto: McClelland and Stewart.

Bakker, Bert, and Mattieu Karel. 1983. "Self-help: Wolf or Lamb?" In *Rediscovering Self-Help: Its Role in Social Care,* edited by D. L. Pancoast, P. Parker, and C. Froland, 159–81. Beverly Hills: Sage Publications.

Bandura, A. 1977a. "Self-Efficacy: Toward a Unifying Theory of Behavioural Change." *Psychological Review* 84(2):191–215.

Bandura, A. 1977b. *Social Learning Theory.* Englewood Cliffs, N.J.: Prentice-Hall.

Bargal, D., and B. Shamir. 1982. "Occupational Welfare as an Aspect of O.W.L." *Labour and Society* 7(3).

Barrera, Manuel Jr. 1981. "Social Support in the Adjustment of Pregnant Adolescents: Assessment Issues." In *Social Networks and Social Support,* edited by B. H. Gottlieb, 69–96. London: Sage Publications.

Barth, P. S., and H. A. Hunt. 1980. *Workers' Compensation and Work-related Illnesses and Diseases.* Cambridge, Mass.: MIT Press.

Beck, K. H. 1983. "Perceived Risk and Risk Acceptance: Implications for Personal

Efficacy." In *Health Risk Estimation, Risk Reduction and Health Promotion,* edited by F. Landry. Papers presented at the 18th Annual Meeting of the Society of Prospective Medicine, Quebec City, Oct. 20–23, 1982. Canadian Public Health Assn., Ottawa, Canada.

Beck, K. H., and A. K. Lund. 1981. "The Effects of Health Threat Seriousness and Personal Efficacy upon Intentions and Behaviour." *Journal of Applied Social Psychology* 11(5):401–15.

Bell, D. 1970. *Work and Its Discontents.* New York: League for Industrial Democracy.

Belle, D. 1982. "Social Ties and Social Support." In *Lives in Stress: Women and Depression,* edited by D. Belle, 133–44. Beverly Hills, Calif.: Sage.

Belloc, N. B., and L. Breslow. 1972. "Relationship of Physical Health Status and Health Practices." *Preventive Medicine* 1:409–21.

Belloc, N. B., L. Breslow, and J. R. Hochstim. 1971. "Measurement of Physical Health in a General Population Survey." *American Journal of Epidemiology* 93(5):328–36.

Benfari, R. C. 1981. "The Multiple Risk Factor Intervention Trial: III. The Model for Intervention." *Preventive Medicine* 10:426–42.

Benson, H. 1975. *The Relaxation Response.* New York: Morrow.

Bergey, E. 1983. Personal communication. Ontario Secondary School Teachers Federation, District 10.

Berkman, L. F., and L. Breslow. 1984. *Health and Ways of Living: The Alameda County Study.* New York: Oxford University Press.

Bird, F. E., and R. G. Loftus. 1976. *Loss Control Management.* Loganville, Georgia: Institute Press.

Blythe, B. J. 1983. "Social Support Networks in Health Care and Health Promotion." In *Social Support Networks: Informal Helping in the Human Services,* edited by J. K. Whittaker et al., 107–31. New York: Aldine.

Borthwick, R. G. 1977. *Summary of Cost Benefit Study Results for Navy Alcoholism Rehabilitation Programs.* Technical Report #346, Arlington, Va.: Presearch Inc., (July).

Bradlyn, A. S., D. P. Strickler, and W. A. Maxwell. 1981. "Alcohol, Expectancy and Stress: Methodological Concerns with the Expectancy Design." *Addictive Behaviors* 6:1–8.

Brandes, S. D. 1976. *American Welfare Capitalism 1880–1940.* Chicago and London: University of Chicago Press.

Brenner, M. H., A. Mooney, and T. J. Nagy, eds. 1980. *Assessing the Contribution of Social Sciences to Health.* AAAS Symposium 26, Boulder, Colo.: Westview Press.

Breslow, L., and J. E. Enstrom. 1980. "Persistence of Health Habits and Their Relationship to Mortality." *Preventive Medicine* 9:469–83.

Briar, K. H., and M. Vinet, 1985. "Ethical Questions Concerning an E.A.P.: Who Is the Client? (Company or Individual?)" In *The Human Resources Management Handbook,* edited by S. H. Klarreich et al. New York: Praeger.

Brown, E. R., and G. E. Margo. 1978. "Health Education: Can the Reformers Be Reformed?" *International Journal of Health Services* 8(1):3–26.

Brown, G. W., M. N. Bhrolchain, and T. Harris. 1975. "Social Class and Psychiatric Disturbance among Women in an Urban Population." *Sociology* 9(2):225–54.

Buck, V. E. 1972. *Working Under Pressure*. New York: Crane, Russak.

Cadbury, E. 1912. *Experiments in Industrial Organization*. London: Longmans, Green.

Cahalan, D., and R. Room. 1974. *Problem Drinking among American Men*. Monograph #7. New Brunswick, N.J.: Rutgers Center of Alcohol Studies.

Caplan, R. D. 1971. "Organizational Stress and Individual Strain: A Social Psychological Study of Risk Factors in Coronary Heart Disease among Administrators, Engineers, and Scientists." Ph.D. diss., University of Michigan (cited in J. S. House, and J. A. Wells, "Occupational Stress, Social Support and Health." In *Reducing Occupational Stress: Proceedings of a Conference*, edited by A. McLean, G. Black, and M. Colligan. Washington, D.C.: US DHEW (NIOSH) Publication #78-140, (April, 1978)).

Cappell, H. 1975. "An Evaluation of Tension Reduction Models of Alcohol Consumption." In *Research Advances in Alcohol and Drug Problems*, edited by R. J. Gibbons et al., 177–209. New York: John Wiley & Sons.

Carey, P. 1985. "Relationship of EAP to Allied Strategies." Paper prepared for EAP Research Conference, Belmont Conference Center. Elkridge, Md. March 31–April 2. N.I.A.A.A. Grant #RI3AA 05906-01.

Carrington, P., et al. 1980. "The Use of Meditation-Relaxation Techniques for the Management of Stress in a Working Population." *Journal of Occupational Medicine* 22(4):221–31.

Cascio, W. F. 1982. *Costing Human Resources: The Financial Impact of Behavior in Organizations*. Boston: Kent.

Cassel, John. 1975. "Social Science in Epidemiology: Psycho-social Processes and 'Stress' Theoretical Formulation." In *Handbook of Evaluation Research*, Vol. 1, edited by E. L. Struening and M. Guttentag, 537–49. Beverly Hills, Calif.: Sage.

Cheek, F. E., and M. DiStefano Miller. 1982. *Prisoners of Life*. American Federation of State, County and Municipal Employees, AFL–CIO. Washington, D.C.

Cherns, Albert. 1978. "The Principles of Sociotechnical Design." In *Sociotechnical Systems: A Sourcebook*, edited by W. A. Pasmore and J. J. Sherwood, 61–71. La Jolla, Calif.: University Associates.

Clemmer, D. 1958. *The Prison Community*. New York: Rinehart.

Cloward, R. A., et al., eds. 1960. *Theoretical Studies in the Social Organization of the Prison*. New York, New York: Conference Group on Correctional Organization. Social Science Research Council.

Cobb, S. 1973. "Role Responsibility: The Differentiation of a Concept." *Occupational Mental Health* 3(1):10–14.

Cobb, S. 1976. "Social Support as a Moderator of Life Stress." *Psychosomatic Medicine* 38(5):300–14.

Compendium of Pharmaceuticals and Specialties, 19th ed. 1984. Ottawa: Canadian Pharmaceutical Assn.

Connelly, P. 1978. *Last Hired, First Fired: Women and the Canadian Work Force*. Toronto: The Women's Press.

Cook. T. D., and D. T. Campbell. 1979. *Quasi-experimentation. Design and Analysis Issues for Field Settings*. Chicago: Rand McNally.

Coolsen, P. 1983. *Strengthening Families through the Workplace*. Chicago, Ill.: National Committee for Prevention of Child Abuse.

Cox, N., R. J. Shephard, and R. Corey. 1981. "Influence of an Employee Fitness Programme upon Fitness Productivity and Absenteeism." *Ergonomics* 24:795–806.

Crawford, R. 1977. "You Are Dangerous to Your Health: The Ideology and Politics of Victim-blaming." *International Journal of Health Services* 7(4):663–80.

Crawford, R. 1980. "Healthism and the Medicalization of Everyday Life." *International Journal of Health Services* 10(3):365–87.

Crouter, Ann C., and J. Garbarino. 1982. "Corporate Self-Reliance and the Sustainable Society." *Technological Forecasting and Social Change* 22:139–51.

Cruickshank, P. J. 1982. "Patient Stress and the Computer in the Consulting Room." *Social Science and Medicine* 16:1371–76.

Cummings, L. L., and T. A. DeCotiis. 1973. "Organizational Correlates of Perceived Stress in a Professional Organization." *Public Personnel Management* (July–August): 275–82.

Cunningham, A. J. 1985. "The Influence of Mind on Cancer." *Canadian Psychology* 26(1):13–29.

Danaher, B. G. 1980. "Smoking Cessation Programs in Occupational Settings." *Public Health Reports* (March–April) 95(2):149–57.

Danielson, D., and K. Danielson. 1980. *Ontario Employee Programme Survey.* Toronto, Ontario: Ministry of Culture and Recreation of Ontario.

Davis, L. E. 1979. "Optimizing Organization-Plant Design: A Complementary Structure for Technical and Social Systems." *Organizational Dynamics* (Autumn): 3–15.

Dean, P. J., L. Reid, and A. Gzowski. 1984. *A Planner's Guide to Fitness in the Workplace.* Toronto, Ontario: Ontario Ministry of Tourism and Recreation.

Derogatis, L. R. 1977. *The SCL-90-R Manual 1: Administration, Scoring, Procedures.* Baltimore, Md.: Clinical Psychometrics Research Unit, Johns Hopkins University School of Medicine.

Derogatis, L. R., and P. A. Cleary. 1977. "Confirmation of the Dimensional Structure of the SCL-90: A Study in Construct Validation." *Journal of Clinical Psychology* 33(4):981–98.

Derogatis, L. R., et al. 1971. "Neurotic Symptom Dimensions as Perceived by Psychiatrists and Patients of Various Social Classes." *Archives of General Psychiatry* 24:454–64.

Derogatis, L. R., et al. 1972. "Factorial Invariance of Symptom Dimensions in Anxious and Depressive Neuroses." *Archives of General Psychiatry* 27:659–65.

Derogatis, L. R., R. S. Lipman, and L. Covi. 1973. "SCL-90: An Outpatient Psychiatric Rating Scale—Preliminary Report." *Psychopharmacology Bulletin* 9(1):13–28.

Derogatis, L. R., K. Rickels, and A. F. Rock. 1976. "The SCL-90 and the MMPI: A Step in the Validation of a New Self-report Scale." *British Journal of Psychiatry* 128:280–89.

Dohrenwend, B. S., and B. P. Dohrenwend, eds. 1974. *Stressful Life Events: Their Nature and Effects.* New York: Wiley-Interscience.

Donovan, D. M., and G. A. Marlatt. 1980. "Assessment of Expectancies and Behaviors Associated with Alcohol Consumption. A Cognitive-Behavioral Approach." *Journal of Studies on Alcohol* 41(11):1153–85.

Dunham, J. 1976. "Stress Situations and Responses." In *Stress in Schools*. National Assoc. of Schoolmasters and Union of Women Teachers. Birmingham–London: Kings Norton Press Group.

Dunn, H. L. 1959. "High Level Wellness for Man and Society." *American Journal of Public Health* 49:286–92.

Eddy, Christen C. 1979. "The Effects of Alcohol on Anxiety in Problem- and Non-problem- Drinking Women." *Alcoholism: Clinical and Experimental Research* 3(2):107–14.

Edwards, G. et al. 1977. "Alcoholism: A Controlled Trial of 'Treatment' and 'Advice.'" *Journal of Studies on Alcohol* 38(5):1004–31.

Eisenstat, R. A., and R. D. Felner. 1984. "Toward a Differentiated View of Burnout: Personal and Organizational Mediators of Job Satisfaction and Stress." *American Journal of Community Psychology* 12(4):411–30.

Ennis, P. 1977. *A Review of the Mortimer–Filkins Test for Identifying Problem Drinkers*. Substudy #854. Toronto, Ontario: Addiction Research Foundation of Ontario.

Ennis, P., and E. Vingilis. 1981. "The Validity of a Revised Version of the Mortimer–Filkins Test with Impaired Drivers in Oshawa, Ontario." *Journal of Studies on Alcohol* 42(7):685–88.

Epstein, S. S. 1979. *The Politics of Cancer*. Garden City, N.J.: Anchor Press.

Erfurt, J. C., and A. Foote. 1984. "Cost-Effectiveness of Work-Site Blood-Pressure Control Programs." *Journal of Occupational Medicine* 26(12):892–900.

Farquhar, J. W., et al. 1977. "Community Education for Cardiovascular Health." *The Lancet* (June 4): 1192–95.

Faust, H. S., and D. Vilnius. 1983. "The Go to Health Project: Summary of Findings with Emphasis on the Relationship between Weight and Absenteeism." In *Health Risk Estimation, Risk Reduction and Health Promotion*, edited by F. Landry. Proceedings of the 18th Annual Meeting, Society of Prospective Medicine, Quebec City. Ottawa: Canadian Public Health Assn.

Fenn, D. 1985. "The Kids Are All Right." *INC* 7(1):48–54.

Fenoaltea, S. 1975. "The Rise and Fall of a Theoretical System: The Manorial System." *Journal of Economic History* 35:386–409.

Ferraro, K. F., E. Mutran, and C. M. Barresi. 1984. "Widowhood, Health and Friendship Support in Later Life." *Journal of Health and Social Behavior* 25:245–59.

Fielding, J. E. 1982. "Effectiveness of Employee Health Improvement Programs." *Journal of Occupational Medicine* 24(11):907–16.

Fielding, J. E., and L. Breslow. 1983. "Health Promotion Programs Sponsored by California Employers." *American Journal of Public Health* 73(5):538–42.

Fields, S. 1980. "Women, Poverty, Depression: Blues in the Red." *Innovations* 7(3):2–19.

Finkel, A. 1977. "Origins of the Welfare State." In *The Canadian State*, edited by L. Panitch. Toronto: University of Toronto Press.

Fishbein, M. 1983. "Factors Influencing Health Behaviours: An Analysis Based on a Theory of Reasoned Action." In *Health Risk Estimation, Risk Reduction and Health Promotion*, edited by F. Landry. Papers presented at the 18th Annual Meeting of the Society of Prospective Medicine, Quebec City, Oct. 20–23, 1982. Ottawa: Canadian Public Health Assn.

Fitch, M. 1982. "Industrial Lung Cancer." *Canadian Family Physician* 28:1787–90.

Foote, A., and J. C. Erfurt. 1983. "Hypertension Control at the Worksite: Comparison of Screening and Referral Alone, Referral and Follow-up and On-Site Treatment." *New England Journal of Medicine* 308:809–813.

Foote, A., et al. 1978. *Cost-effectiveness of Occupational Employee Assistance Programs*. Ann Arbor, Mich.: Worker Health Program, Institute of Labour and Industrial Relations. University of Michigan–Wayne State University.

Francek, J. L. 1985. "The Role of the Occupational Social Worker in EAPs." In *The Human Resources Management Handbook,* edited by S. H. Klarreich, J. L. Francek, and C. E. Moore. New York: Praeger.

Francek, J. L., S. H. Klarreich, and C. E. Moore. 1985. "The Future of EAPs and New Directions." In *The Human Resources Management Handbook,* edited by S. H. Klarreich, J. L. Francek, and C. E. Moore. New York: Praeger.

Freeman, V. J. 1960. "Beyond Germ Theory: Human Aspects of Health and Illness." *Journal of Health and Human Behaviour* 1(1):8–13.

Freire, P. 1971. *Pedagogy of the Oppressed*. New York: Seabury Press.

Freire, P. 1973. *Education for Critical Consciousness*. New York: Seabury Press.

French, J. R. P. Jr., R. D. Caplan, and R. Van Harrison, 1982. *The Mechanisms of Job Stress and Strain*. Toronto: John Wiley & Sons.

Friedman, G. H., B. E. Lehrer, and J. P. Stevens. 1983. "The Effectiveness of Self-directed and Lecture/Discussion Stress Management Approaches and the Locus of Control of Teachers." *American Educational Research Journal* 20(4):563–80.

Froland, C., et al. 1981. "Linking Formal and Informal Support Systems." In *Social Networks and Social Support,* edited by B. H. Gottlieb, 259–75. London: Sage.

Frydman, M. I. 1981. "Social Support, Life Events and Psychiatric Symptoms: A Study of Direct, Conditional and Interaction Effects." *Social Psychiatry* 16:69–78.

Galdston, I. 1954. *Beyond Germ Theory*. New York: Health Education Council.

Gelderman, C. 1981. *Henry Ford, the Wayward Capitalist*. New York: Dial Press.

Giesbrecht, N., G. Markle, and S. MacDonald. 1982. "The 1978–1979 INCO Workers' Strike in the Sudbury Basin and Its Impact on Alcohol Consumption and Drinking Patterns." *Journal of Public Health Policy* 3(1):22–38.

Glaser, Edward M. 1974. *Improving the Quality of Worklife . . . and in the Process, Improving Productivity. A Summary of Concepts, Procedures and Problems, with Case Histories*. Los Angeles: Human Interaction Research Inst.

Glossop, Robert. 1980. "Research and the Dilemmas of Action." For presentation at the 20th Anniversary Symposium on "The Family" of the Children's Psychiatric Research Institute, May 22–23, London, Ontario.

Goetz, A. A., J. F. Duff, and M. D. Bernstein. 1980. "Health Risk Appraisal: The Estimation of Risk." *Public Health Reports* 95(2):119–26.

Goffman, E. 1961. *Asylums*. Garden City, New York: Anchor Books.

Gore, Susan. 1978. "The Effects of Social Support in Moderating the Health Consequences of Unemployment." *Journal of Health and Social Behavior* 19:157–65.

Gottlieb, B. H. 1981. "Social Networks and Social Support in Community Mental Health." In *Social Networks and Social Support,* edited by B. H. Gottlieb, 11–42. London: Sage.

Gottlieb, B. H. 1982. "Social Support in the Workplace." In *Community Support Systems and Mental Health, Practice, Policy and Research,* edited by D. E. Biegel and A. J. Naparstek, 37–53. New York: Springer.

Gray, R., and L. M. Poudrier. 1982. "The Prevalence of Problem Drinking among Government Employees Four Years after Introducing an E.A.P." Unpublished manuscript. Toronto, Ontario: Addiction Research Foundation.

Groen, J. J., and J. Bastiaans. 1975. "Psychosocial Stress, Interhuman Communication and Psychosomatic Disease." In *Stress and Anxiety,* Vol. 1, edited by C. D. Spielberger and I. G. Sarason, 27–49. Washington, D.C.: Hemisphere.

Groeneveld, J., et al. 1984. *The Alcoholism Treatment Program at Canadian National Railways.* Toronto, Ont.: Addiction Research Foundation Working Paper Series.

Groeneveld, J., et al. 1985. "Job retention rates for program planning." *EAP Digest* (Jan.–Feb.): 29–38.

Gruenberg, B. 1980. "The Happy Worker: An Analysis of Education and Occupational Differences in Determinants of Job Satisfaction." *American Journal of Sociology* 86(2):247–71.

Guest, R. H. 1979. "Quality of Work Life—Learning from Tarrytown." *Harvard Business Review* (July–August): 76–87.

Hackman, J. R., and M. D. Lee. 1979. *Redesigning Work: A Strategy for Change.* Work in America Institute Studies in Productivity, 9. Scarsdale, N.Y.: Work in America Inst. Inc.

Hammen, C., and A. Mayol. 1982. "Depression and Cognitive Characteristics of Stressful Life-Event Types." *Journal of Abnormal Psychology* 91(3):165–74.

Hammer, M. 1981. "Social Supports, Social Networks, and Schizophrenia." *Schizophrenia Bulletin* 7(1):45–57.

Hareven, T. K., and R. Langenbach. 1978. *Amoskeag.* New York: Pantheon Books.

Haw, M. A. 1982. "Women, Work and Stress: A Review and Agenda for the Future." *Journal of Health and Social Behavior* 23:132–44.

Haynes, S. G., and M. Feinleib. 1980. "Women, Work and Coronary Heart Disease: Prospective Findings from the Framingham Heart Study." *American Journal of Public Health* 70:133–41.

Hays, S. P. 1957. *The Response to Industrialism 1885–1914.* Chicago: University of Chicago Press.

Henderson, S. 1980. "A Development in Social Psychiatry: The Systematic Study of Social Bonds." *The Journal of Nervous and Mental Disease* 168(2):63–69.

Herbert, C. P., and G. M. Gutman. 1980. "Practical Group Autogenic Training for Management of Stress-related Disorders in Family Practice." In *Clinical Hypnosis in Medicine,* edited by H. Wain, 109–18. Chicago: Year Book Medical Publishers.

Herzberg, F. 1967. *Work and the Nature of Man.* New York: World Publishing Company.

Higgins, R. L., and G. A. Marlatt. 1975. "Fear of Interpersonal Evaluation as a Determinant of Alcohol Consumption in Male Social Drinkers." *Journal of Abnormal Psychology* 84(6):644–51.

Hillenberg, J. B., and F. L. Collins. 1982. "A Procedural Analysis and Review of Relaxation Training Research." *Behaviour Research & Therapy* 20:251–60.

Hillenberg, J. B., and F. L. Collins, 1983. "The Importance of Home Practice for Progressive Relaxation Training." *Behaviour Research & Therapy* 21(6):633–42.

Hinrichs, J. R. 1978. *Practical Management for Productivity*. New York: Van Nostrand Reinhold.

Hire, J. N. 1978. "Anxiety and Caffeine." *Psychological Reports* 42(3):833–34.

Hitchcock, L. C., and M. Sanders. 1976. *A Survey of Alcohol and Drug Abuse Programs in the Railroad Industry*. National Technical Information Service. Springfield, VA: U.S. Dept. of Commerce.

Hitz, D. 1973. "Drunken Sailors and Others: Drinking Problems in Specific Occupations." *Quarterly Journal of Studies in Alcohol* 34:496–505.

Holmes, T. H., and R. H. Rahe. 1967. "The Social Readjustment Rating Scale." *Journal of Psychosomatic Research* 11:213–18.

House, J. S. 1974. "Occupational Stress and Coronary Heart Disease: A Review and Theoretical Integration." *Journal of Health and Social Behavior* 15:12–27.

House, J. S., J. M. LaRocco, and J. R. P. French. 1982. "Response to Schaefer." *Journal of Health and Social Behavior* 23:98–101.

House, J. S., and J. A. Wells. 1978. "Occupational Stress, Social Support and Health." In *Reducing Occupational Stress: Proceedings of a Conference*, edited by A. McLean, G. Black, and M. Colligan, 8–29. Washington, D.C.: U.S. DHEW (NIOSH) Publication No. 78-140.

Ilfeld, F. W. 1976. "Further Validation of a Psychiatric Symptom Index in a Normal Population." *Psychological Reports* 39:1215–28.

Ison, T. G. 1983. *Workers' Compensation in Canada*. Toronto: Butterworths.

Jacobsen, G. 1975. *Diagnosis and Assessment of Alcohol Abuse and Alcoholism*. Rockville, MD: National Institute of Alcohol Abuse and Alcoholism.

Janis, I. L., and L. Mann. 1977. *Decision-making: A Psychological Analysis of Conflict, Choice and Commitment*. New York: Free Press.

Johnson, J. H., and Sarason, I. G. 1978. "Life Stress, Depression and Anxiety: Internal-External Control as a Moderator Variable." *Journal of Psychosomatic Research* 22:205–8.

Josephson, H. 1949, 1967. *The Golden Threads*. New York: Russell and Russell.

Kahn, R. L., et al. 1964. *Organizational Stress: Studies in Role Conflict and Ambiguity*. New York: John Wiley & Sons.

Kasl, S. V. 1978. "Epidemiological Contributions to the Study of Work Stress." In *Stress at Work*, edited by C. L. Cooper and R. Payne, 3–48. New York: John Wiley and Sons.

Katz, A. H., and L. S. Levin. 1980. "Self-Care Is Not a Solipsistic Trap: A Reply to Critics." *International Journal of Health Services* 10(2):329–36.

Katzell, R. A., P. Bienstock, and P. H. Faerstein. 1977. *A Guide to Worker Productivity Experiments in the United States, 1971–75*. New York: New York University Press.

Katzell, R. A., et al. 1975. *Work, Productivity and Job Satisfaction. An Evaluation of Policy-Related Research*. The Psychological Corporation.

Kerlans, M., et al. 1971. *Court Procedures for Identifying Problem Drinkers. Vol. 1: Manual*. N.H.T.S.A. Report #DOT-HS-800-632. Ann Arbor, Mich.: Highway Safety Research Institute. University of Michigan.

Klos, D. M., and I. M. Rosenstock. 1982. "Some Lessons from the North Karelia Project." *American Journal of Public Health* 72(1):53–54.

Kohn, M. L. 1976. "Occupational Structure and Alienation." *American Journal of Sociology* 82(1):111–30.

Kohn, M. L., and C. Schooler. 1973. "Occupational Experience and Psychological Functioning: An Assessment of Reciprocal Effects." *American Sociological Review* 38:97–118.

Kok, F. J., et al. 1982. "Characteristics of Individuals with Multiple Behavioral Risk Factors for Coronary Heart Disease: The Netherlands." *American Journal of Public Health* 72(9):986–91.

Kornhauser, A. 1965. *Mental Health of the Industrial Worker: A Detroit Study.* New York: John Wiley & Sons.

Kornitzer, M. D., et al. 1981. "Workload and Coronary Heart Disease." In *Myocardial Infarction and Psychosocial Risks,* edited by J. Siegrist and M. J. Halhuber, 18–40. Berlin: Springer.

Kristein, M. M. 1982. "The Economics of Health Promotion at the Worksite." *Health Education Quarterly* (Special Supplement) 9:27–36.

Kromhout, D., E. B. Bosscheiter, and C. de Lezenne Coulandes. 1982. "Dietary Fibre and 10-Year Mortality from Coronary Heart Disease, Cancer and All Causes: The Zutphen Study." *The Lancet* (Sept. 4): 518.

LaLonde, M. 1974. *A New Perspective on the Health of Canadians.* Ottawa: National Health and Welfare.

La Rocco, F. M., J. S. House, and J. R. P. French. 1980. "Social Support, Occupational Stress, and Health." *Journal of Health and Social Behavior* 21(3):202–18.

Labonté, R. N., and S. P. Penfold. 1981. *Health Promotion Philosophy: From Victim-blaming to Social Responsibility.* Vancouver: Ronald Labonte.

Landry, F., ed. 1983. *Health Risk Estimation, Risk Reduction and Health Promotion.* Papers presented at the 18th Annual Meeting of the Society of Prospective Medicine, Quebec City, Oct. 20–23, 1982. Ottawa: Canadian Public Health Assn.

Larbi, E. B., et al. 1984. "The Population Attributable Risk of Hypertension from Heavy Alcohol Consumption." *Public Health Reports* 99(3):316–19.

Lavoie, F. 1983. "Citizen Participation in Health Care." In *Rediscovering Self-Help: Its Role in Social Care,* edited by D. L. Pancoast, P. Parker, and C. Froland, 225–38. Beverly Hills, Calif.: Sage.

Lehrer, P. M., et al. 1983. "Progressive Relaxation and Meditation. A Study of Psychophysiological and Therapeutic Differences between Two Techniques." *Behaviour Research & Therapy* 21(6):651–62.

Lesser, P. J. 1967. "The Legal Viewpoint." In *To Work Is Human: Mental Health and the Business Community,* edited by A. McLean, 103–22. New York: Macmillan.

Lesser, Philip, and Ari Kiev. 1970. "Psychiatric Disability and Workmen's Compensation." In *Mental Health and Work Organizations,* edited by A. McLean, 237–50. Chicago: Rand McNally.

Levinson, H., and L. Weinbaum. 1970. "The Impact of Organization on Mental Health." In *Mental Health and Work Organizations,* edited by A. McLean, 23–49. Chicago: Rand McNally.

Lin, N., A. Dean, and W. M. Ensel. 1981. "Social Support Scale Methodological Note." *Schizophrenia Bulletin* 7(1):73–88.

Lipscomb, T., et al. 1980. "Effects of Tolerance on the Anxiety-Reducing Function of Alcohol." *Archives of General Psychiatry* 37:577–82.

Logue, P. E., et al. 1978. "Effects of Alcohol Consumption on State Anxiety Changes in Male and Female Nonalcoholics." *American Journal of Psychiatry* 135(9):1079–81.

Lowenthal, M. F. 1964. "Social Isolation and Mental Illness in Old Age." *American Sociological Review* 29:54–70. (Cited by B. Cooper and U. Sosna, "Family Settings of the Psychiatrically Disturbed Aged." In *The Social Consequences of Psychiatric Illness*, L. N. Robins et al., 141–57. New York: Brunner/Mazel, 1980).

Lumsden, D. P. 1981. "Is the Concept of Stress of Any Use Anymore?" In *Contributions to Primary Prevention in Mental Health*, edited by D. Randall. Toronto: Canadian Mental Health Assn.

McAlister, A. L. 1983. "Factors Influencing Decisions to Modify Health Behaviour." In *Health Risk Estimation, Risk Reduction and Health Promotion*, edited by F. Landry. Papers presented at the 18th Annual Meeting of the Society of Prospective Medicine, Quebec City, Oct. 20–23, 1982. Canadian Public Health Assn., Ottawa, Canada.

McAlister, A. et al., 1982. "Theory and Action for Health Promotion: Illustrations from the North Karelia Project." *American Journal of Public Health* 72(1):43–50.

MacBride, A., et al. 1981. "Occupational Stress among Canadian Air Transportation Administration Employees: Ontario Region." Toronto, Ont.: Clarke Institute of Psychiatry, Preliminary Report, (March).

McCorkle, L. W., and R. Korn. 1954. "Resocialization within walls." *Annals of the American Academy of Political and Social Science* 293, (May).

McFarlane, A. H., et al. 1983. "The Process of Social Stress: Stable, Reciprocal and Mediating Relationships." *Journal of Health and Social Behavior* 24:160–73.

McLean, A. 1974. "Concepts of Occupational Stress, A Review." In *Occupational Stress*, edited by A. McLean, 3–14. Springfield, Ill.: Charles C. Thomas.

McMichael, A. J. 1978. "Personality, Behavioural and Situational Modifiers of Work Stressors." In *Stress at Work*, edited by C. L. Cooper and R. Payne, 127–47. Toronto: J. Wiley and Sons.

McQueen, D. V., and D. D. Celentano. 1982. "Social Factors in the Etiology of Multiple Outcomes: The Case of Blood Pressure and Alcohol Consumption Patterns." *Social Science and Medicine* 16:397–418.

McQueen, D. V., and J. Siegrist. 1982. "Social Factors in the Etiology of Chronic Disease: An Overview." *Social Science and Medicine* 16:353–67.

Mackie, L., and P. Patullo. 1977. *Women at Work*. London: Tavistock.

Makosky, V. P. 1982. "Sources of Stress: Events or Conditions?" In *Lives in Stress: Women and Depression*, edited by D. Belle, 35–53. Beverly Hills, Calif.: Sage.

Mannello, R. A., and F. J. Seaman. 1979. *Prevalence, Costs and Handling of Drinking Problems on Seven Railroads. Final Report*. Washington, D.C.: University Research Corporation.

Mansfield, P. K. 1982. "Women and Work: A Proposal for Women's Occupational Health Education." *Health Education* (Sept./Oct.): 5–8.

Manuso, J. 1981. "Psychological Services and Health Enhancement: A Corporate Model." In *Linking Health and Mental Health: Coordinating Care in the Community. Sage Annual Review of Mental Health,* vol. 2. Edited by Anthony Broskowski et al. Beverly Hills, Calif.: Sage.

Margolis, B., and W. H. Kroes. 1973. "Occupational Stress and Strain." *Occupational Mental Health* 2(4):4–6.

Margolis, B., W. H. Kroes, and R. P. Quinn. 1974. "Job Stress: An Unlisted Occupational Hazard." *Journal of Occupational Medicine* 16(10):659–61.

Marlatt, G. A. 1976. "Alcohol, Stress and Cognitive Control." In *Stress and Anxiety,* vol. 3, edited by I. G. Sarason and C. D. Spielberger. Washington, D.C.: Hemisphere.

Marlatt, G. A. 1979. "Alcohol Use and Problem Drinking: A Cognitive-Behavioral Analysis." In *Cognitive-Behavioral Interventions,* edited by P. C. Kendall and S. D. Hollon, 319–55. New York: Academic Press.

Marlatt, G. A., and J. R. Gordon. 1980. "Determinants of Relapse: Implications for the Maintenance of Behavior Change." In *Behavioral Medicine: Changing Health Lifestyles,* edited by P. O. Davidson and S. M. Davidson, 410–52. New York: Brunner/Mazel.

Marlatt, G. A., C. F. Kosturn, and A. R. Lang. 1975. "Provocation to Anger and Opportunity for Retaliation as Determinants of Alcohol Consumption in Social Drinkers." *Journal of Abnormal Psychology* 34:652–59.

Marlatt, G. A., and J. K. Marques. 1977. "Meditation, Self-Control and Alcohol Use." In *Behavioral Self-Management: Strategies, Techniques and Outcomes,* edited by R. B. Stuart, 117–53. New York: Brunner/Mazel.

Marmot, M. G., et al. 1981. "Alcohol and Mortality: A U-shaped Curve." *The Lancet* (March 14): 580–83.

Marx, K. 1964. *Early Writings.* Edited and translated by T. B. Bottomore. New York: McGraw Hill.

Masi, D. 1982. *Human Services in Industry.* Lexington, MA: Lexington Books, D.C. Heath Co.

Mechanic, D., and P. D. Cleary. 1980. "Factors Associated with the Maintenance of Positive Health Behavior." *Preventive Medicine* 9:805–14.

Merwin, D. J., and B. A. Northrop. 1982. "Health Action in the Workplace: Complex Issues—No Simple Answers." *Health Education Quarterly,* Vol. 9, Special Supp.

Miller, R. E., M. Shain, and T. J. Golaszewski. 1985. *The Synergism of Health Promotion and Restoration in the Prevention of Substance Abuse in the Workplace. Health Values: Achieving High Level Wellness.* In press.

Miller, W. M. 1980. *The Addictive Behaviors.* Elmsford, N.Y.: Pergamon Press.

Minkler, M., and K. Cox. 1980. "Creating Critical Consciousness in Health: Applications of Freire's Philosophy and Methods to the Health Care Setting." *International Journal of Health Services* 10(2):311–23.

Mortimer, R. G., et al. 1971. *Court Procedures for Identifying Problem Drinkers; Report on Phase 1.* Interim Report #DOT-HS-800-630. Ann Arbor, Mich.: University of Michigan, Highway Safety Research Institute.

Mueller, D. P. 1980. "Social Networks: A Promising Direction for Research on the Relationship of the Social Environment to Psychiatric Disorder." *Social Science and Medicine* 14:147–61.

Mules, J. E., W. H. Hague, and D. L. Dudley. 1977. "Life Change, Its Perception and Alcohol Addiction." *Journal of Studies on Alcohol* 38(3):487–93.

Multiple Risk Factor Intervention Trial Research Group. 1982. "Multiple Risk Factor Intervention Trial. Risk factor changes and mortality results." *Journal of the American Medical Association* 248(12):1465–77.

Navarro, V. 1976. "Social Class, Political Power and the State and Their Implications in Medicine." *Social Science and Medicine* 10:437–57.

North, D., and R. Thomas. 1971. "The Rise and Fall of the Manorial System: A Theoretical Model." *Journal of Economic History* 31 (Dec.):777–803.

Nye, D. E. 1979. *Henry Ford—Ignorant Idealist.* Port Washington, New York: Kennikat Press.

Oliver, M. F. 1982. "Does Control of Risk Factors Prevent Coronary Heart Disease?" *British Medical Journal,* 285 (Oct.):1065–66.

Orford, J., E. Oppenheimer, and G. Edwards, 1976. "Abstinence or Control: The Outcome for Excessive Drinkers Two Years after Consultation." *Behaviour Research & Therapy* 14:409–18.

Orme-Johnson, D. W., and J. T. Farrow, eds. 1978. *Scientific Research on the Transcendental Meditation Programme,* vol. 1. Livingston Manor, N.Y.: Maharishi European Research University Press.

Orvis, B. R., et al. 1981. *Effectiveness and Cost of Alcohol Rehabilitation in the U.S. Air Force.* Santa Monica: Rand.

Ozawa, M. 1980. "Development of Social Services in Industry: Why and How?" *Social Work* 25 (Nov.): 464–70.

Parker, D. A., and J. A. Brody. 1982. "Risk Factors for Occupational Alcoholism and Alcohol Problems." In *Occupational Alcoholism: A Review of Research Issues.* Research Monograph #8. Rockville, Md.: National Institute on Alcohol Abuse and Alcoholism.

Parkinson, R. S., et al., eds. 1982. *Managing Health Promotion in the Workplace: Guidelines for Implementation and Evaluation.* Palo Alto: Mayfield.

Pasmore, W. A., and J. J. Sherwood, eds. 1978. *Sociotechnical Systems: A Sourcebook.* La Jolla, Calif.: University Associates.

Pattison, E. M., M. B. Sobell, and L. C. Sobell, eds. 1977. *Emerging Concepts of Alcohol Dependence.* New York: Springer.

Pearlin, L. I., and C. Radabaugh. 1976. "Economic Strains and Coping Functions of Alcohol." *American Journal of Sociology* 82:652–63.

Pearlin, L. I., and C. Schooler. 1978. "The Structure of Coping." *Journal of Health and Social Behaviour* 19:2–21.

Peters, R. K., H. Benson, and D. Porter. 1977a. "Daily Relaxation-Response Breaks in a Working Population: 1. Effects of Self-Reported Measures on Health, Performance and Well-being." *American Journal of Public Health* 67(10):946–52.

Peters, R. K., H. Benson, and D. Porter. 1977b. "Daily Relaxation Breaks in a Working Population. 2. Effects on Blood Pressure." *American Journal of Public Health* 67(10):954–59.

Peters, T. J., and R. H. Waterman. 1982. *In Search of Excellence.* New York: Warner Books.

Piliavin, I. 1966. "Reduction of Custodian-Professional Conflict in Correctional Institutions." *Crime and Delinquency* 12(2):125–34.

Pinneau, S. R. 1975. "Effects of Social Support on Psychological and Physiological Strain." Ph.D. diss., University of Michigan.

Pinneau, S. R. 1976. "Effects of Social Support on Occupational Stresses and Strain." Paper presented at the 84th Annual Convention of the American Psychological Association, Washington, D.C. (Sept.).

Pinneau, S. R., and A. Newhouse. 1964. "Measures of Invariance and Comparability in Factor Analysis for Fixed Variables." *Psychometrika* 29:271–81.

Plant, Martin A. 1977. "Alcoholism and Occupation: A Review." *British Journal of Addictions* 72:309–16.

Plant, Martin A. 1978. "Occupation and Alcoholism: Cause or Effect? A Controlled Study of Recruits to the Drink Trade." *The International Journal of the Addictions*, 13(4):605–26.

Poole, A. L. 1955. *From Domesday Book to Magna Carta 1087–1216. Oxford History of England 2nd Edition.* Oxford: Clarendon Press.

Popple, P. R. 1981. "Social Work Practice in Business and Industry." *Social Service Review* (June): 257–69.

Porterfield, S. L. 1960. "Social Knowledge in Medicine: An Editorial." *Journal of Health and Human Behavior* 1(1):3–8.

President's Commission on Mental Health. 1978. *Report to the President.* Vols. 1, 2, 4. Washington, D.C.: U.S. Government Printing Office.

Pupo, N., M. Shain, and M. Boutilier. 1985. "Evaluation of a Healthy Lifestyles Course for Women." Unpublished manuscript. Toronto, Ontario: Addiction Research Foundation.

Puska, P., L. Neittaanmaki, and J. Tuomilehto. 1981. "A Survey of Local Health Personnel and Decision-makers Concerning the North Karelia Project: A Community Program for Control of Cardiovascular Diseases." *Preventive Medicine* 10:564–76.

Puska, P., et al. 1979. "Changes in Coronary Risk Factors during a Comprehensive Five-Year Community Programme to Control Cardiovascular Diseases: (North Karelia Project)." *British Medical Journal* 2 (Nov. 10): 1173–78.

Rabkin, J. G., and E. L. Struening. 1976. "Life Events, Stress and Illness." *Science* 194:1013–20.

Rahe, R. 1975. "Epidemiological Studies of Life Change and Illness." *International Journal of Psychiatry in Medicine* 6(1–2):133–46.

Reasons, C. E., L. L. Ross, and C. Paterson. 1981. *Assault on the Worker. Occupational Health and Safety in Canada.* Toronto: Butterworths.

Rinehart, J. W. 1975. *The Tyranny of Work.* Don Mills, Ontario: Longman Canada.

Roemer, M. I. 1984. "The Public/Private Mix of Health Sector Financing: International Implications." *Public Health Reviews* 12(2):119–30.

Rokeach, M. 1973. *The Nature of Human Values.* New York: Free Press.

Rokeach, M. 1979. "Value Theory and Communication Research." In *Communication Yearbook III,* edited by D. Nimmo. New Brunswick, N.J.: Transaction Books.

Roman, P. M. 1980. *Employee Alcoholism Programs in Major Corporations in 1979: Scope, Change, and Receptivity.* Report in the Tulane University Monitoring and Evaluation of Occupational Alcoholism Programming. Tulane University, New Orleans.

Roman, P. M. 1981a. "Evaluation of Employee Alcoholism Programs." *Labor-Management Alcoholism Journal* 10(1):1–12.

Roman, P. 1981b. *Prevention and Health Promotion Programming for Work Organizations: Employee Assistance Program Experience.* Prevention Resources Project Monograph. DeKalb, Illinois: Northern Illinois University.

Roman, P. 1984. Quoted in Editorial "Jury Is Still Out on EAP–Wellness Program Integration." *Employee Health and Fitness* 6(7):77–88, at pgs, 78, 79.

Roman, P. M., and H. M. Trice. 1970. "The Development of Deviant Drinking Behavior: Occupational Risk Factors." *Archives of Environmental Health* 20 (March): 424–35.

Romeder, J. M. 1982. *Self Help Groups in Canada.* Social Service Development and Grants Directorate Branch, Health and Welfare, Ottawa, Ontario, Canada.

Rosen, G., ed. 1949. *Backgrounds in Social Medicine.* New York: Millbank Memorial Fund.

Roskies, E., and R. S. Lazarus. 1980. "Coping Theory and the Teaching of Coping Skills." In *Behavioural Medicine: Changing Health Lifestyles,* edited by P. O. Davidson and S. M. Davidson, 38–69. New York: Brunner/Mazel Publishers.

Salonen, J. T., P. Puska, and H. Mustaniemi. 1979. "Changes in Morbidity and Mortality during Comprehensive Community Programme to Control Cardiovascular Diseases During 1972–1977 in North Karelia." *British Medical Journal* 2:1178–83.

Sanua, V. D. 1960. "Socio-Cultural Factors in Responses to Stressful Life Situations: Aged Amputees as an Example." *Journal of Health and Human Behavior* 1(1):17–25.

Sapolsky, H. M., et al. 1981. "Corporate Attitudes towards Health Care Costs." *Millbank Memorial Fund Quarterly* 59(4):561–85.

Sarason, I. G., J. H. Johnson, and J. M. Siegel. 1978. "Assessing the Impact of Life Changes: Development of the Life Experiences Survey." *Journal of Consulting and Clinical Psychology* 46(5):932–46.

Sarason, I. G., et al. 1983. "Assessing Social Support: The Social Support Questionnaire." *Journal of Personality and Social Psychology* 44(1):127–39.

Schaefer, C., J. C. Coyne, and R. S. Lazarus. 1981. "The Health-Related Functions of Social Support." *Journal of Behavioral Medicine* 4(4):381–406.

Schrag, C. 1954. "Leadership among prison inmates." *American Sociological Review* 19 (Feb.):37–42.

Schramm, C. J. 1982. *Evaluating Occupational Programs: Efficiency and Effectiveness.* In Research Monograph #8. Rockville, Md.: National Institute on Alcohol Abuse and Alcoholism.

Schwartz, G. E. 1982. "Stress Management in Occupational Settings." In *Managing Health Promotion in the Workplace,* edited by R. S. Parkinson et al., 233–51. Palo Alto: Mayfield.

Seeman, M. 1959. "On the Meaning of Alienation." *American Sociological Review* 24:783–91.

Seeman, M., and C. S. Anderson. 1983. "Alienation and Alcohol: The Role of Work, Mastery, and Community in Drinking Behavior." *American Sociological Review* 48:60–77.

Selye, H. 1950. *The Physiology and Pathology of Exposure to Stress*. Montreal: Acta.

Selye, H. 1956. *The Stress of Life*. New York: McGraw-Hill.

Selye, H. 1976. *Stress in Health and Disease*. London: Butterworths.

Shain, M. 1976. "Prerequisites for Evaluative Research in Halfway Houses—Decision-making Aids for Administrator and Researcher." *Evaluation* 3:1–2, 47–50.

Shain, M. 1981. "The Value of Health Promotion in the Prevention of Alcohol and Drug Abuse." Toronto, Ont.: Occupational Programmes Research Section Manuscript, Addiction Research Foundation.

Shain, M. 1985. "An Exploration of the Ability of Broad-based E.A.P.s to Generate Alcohol-related Referrals." In *The Human Resources Management Handbook: Principles and Practice of E.A.P.*, edited by J. H. Klarreich et al. New York: Praeger.

Shain, M., M. Boutilier, and J. Simon. 1984. "Stress, Strain and the Use of Alcohol and Drugs among a Sample of Teachers." Unpublished manuscript. Toronto, Ont.: Occupational Programmes Research, Addiction Research Foundation.

Shain, M., and B. Boyle. 1985. "Toward Coordination of Employee Health and Assistance Programmes." In *The Human Resources Management Handbook: Principles and Practices of E.A.P.s*, edited by J. H. Klarreich, J. L. Francek, and C. E. Moore. New York: Praeger Press.

Shain, M., and J. Groeneveld. 1980. *Employee Assistance Programmes: Philosophy, Theory and Practice*. Lexington, MA: Lexington Books, D.C. Heath Co.

Shapiro, S. 1977. "Measuring the Effectiveness of Prevention: II." *Millbank Memorial Fund Quarterly* (Spring): 291–306.

Sherwin, R., et al. 1981. "The Multiple Risk Factor Intervention Trial (MRFIT): II. The Development of the Protocol." *Preventive Medicine* 10:402–25.

Sidel, V. W. 1979. "Public Health in International Perspective: From 'Helping the Victim' to 'Blaming the Victim' to 'Organizing the Victims.'" *Canadian Journal of Public Health* 70(4):234–39.

Simmons, R., B. J. Kay, and C. Regan. 1984. "Women's Health Groups: Alternatives to the Health Care System." *International Journal of Health Services* 14(4):619–34.

Simpson, M. E. 1980. "Societal Support and Education." In *Handbook on Stress and Anxiety*, edited by I. L. Kutash et al., 451–62. Washington, D.C.: Jossey-Bass.

Skinner, H. A. 1979. "A Multivariate Evaluation of the MAST." *Journal of Studies on Alcohol* 40(9):831–44.

Smart, R. G. 1979. "Drinking Problems Among Employed, Unemployed and Shift Workers." *Journal of Occupational Medicine* 21(11):731–36.

Sobell, M. B., and L. C. Sobell. 1978. *Behavioural Treatment of Alcohol Problems: Individualized Therapy and Controlled Drinking*. New York and London: Plenum Press.

Sonnenstuhl, W. J. 1982. "Understanding EAP Self-referral: Toward a Social Network Approach." *Contemporary Drug Problems* (Summer): 269–93.

Spielberger, C. D. 1975. "Anxiety: State-trait Process." In *Stress and Anxiety*, vol. 1, edited by C. D. Spielberger and I. G. Sarason, 115–43. Washington, D.C.: Hemisphere.

Spielberger, C. D., R. L. Gorsuch, and R. E. Lushene. 1970. *STAI Manual for the*

State-Trait Anxiety Inventory. Palo Alto, Calif.: Consulting Psychologists Press.

Steinert, Y. 1978. "Helping Patients Cope with Stressful Life Events." *Canadian Family Physician* 24:859–62.

Street, D., R. D. Vinter, and C. Perrow. 1966. *Organization for Treatment. A Comparative Study of Institutions for Delinquents.* New York: Free Press.

Suls, J., and B. Mullen. 1981. "Life Change and Psychological Distress: The Role of Perceived Control and Desirability." *Journal of Applied Social Psychology* 11(5):379–89.

Sutton, R. I. 1984. "Job Stress among Primary and Secondary Schoolteachers: Its Relationship to Ill-being." *Work and Occupations* 11(1):7–28.

Suurvali, H., and M. Shain. 1981. *Workplace Innovations: Implications for the Incidence and Management of Problems due to Alcohol and Drug Abuse.* Substudy No. 1176. Toronto, Ont.: Alcoholism and Drug Addiction Research Foundation.

Swint, J. M. 1982. Critique of "Evaluating Occupational Programs: Efficiency and Effectiveness," by C. J. Schramm. In *Research Monograph #8.* Rockville, Md.: National Institute on Alcohol Abuse and Alcoholism.

Sykes, G. M., and S. L. Messinger. 1970. "The Inmate Social Code." In *The Sociology of Punishment and Corrections,* edited by N. Johnston, L. Savitz, and M. Wolfgang, 401–8. New York: John Wiley & Sons.

Tesh, S. 1984. "The Politics of Stress: The Case of Air Traffic Control." *International Journal of Health Services* 14(4):569–87.

Theorell, T., et al. 1975. "The Relationship of Disturbing Life Changes and Emotions to the Early Development of Myocardial Infarction and Other Serious Illnesses." *International Journal of Epidemiology* 4:281–93.

Thoits, P. A. 1982. "Conceptual, Methodological, and Theoretical Problems in Studying Social Support as a Buffer Against Life Stress." *Journal of Health and Social Behavior* 23:145–59.

Thomlison, R. J., ed. 1983. *Perspectives on Industrial Social Work Practice.* Chapter 1. Ottawa: Family Service Canada Publications.

Tolson, A. 1977. *The Limits of Masculinity.* London: Tavistock.

Trice, H. M. 1983. "Employee Assistance Programs. Where Do We Stand in 1983?" *Journal of Psychiatric Treatment and Evaluation* 5:521–29.

Trice, H. M., and J. A. Belasco. 1966. *Emotional Health and Employer Responsibility.* Bulletin 57. Ithaca, N.Y.: New York State School of Industrial and Labor Relations, Cornell University.

Trice, H. M., and J. M. Beyer. 1984. "Employee Assistance Programs: Blending Performance-oriented and Humanitarian Ideologies to Assist Emotionally Disturbed Employees." In *Research in Community and Mental Health,* vol. 4, edited by J. R. Greenley, 245–97. Greenwich, CT: JAI Press.

Trice, H. M., and M. Schonbrunn. 1981. "A History of Job-based Alcoholism Programs: 1900–1955." *Journal of Drug Issues* 11:171–98.

Turner, R. J., B. G. Frankel, and D. Levin. 1982. "Social Support: Conceptualization, Measurement and Implications for Mental Health." Ontario: University of Western Ontario, Health Care Research Unit. (Also in *Research in Community and Mental Health,* vol. 3, edited by James R. Greenley. Greenwich: JAI Press, forthcoming.)

U.S. Department of Health, Education and Welfare. Public Health Service. 1976. *For-*

ward Plan for Health FY 1978–82. Washington, D.C.: U.S. Dept. of Health, Education and Welfare.

U.S. Department of Health, Education and Welfare, Public Health Service. 1979. *Healthy People: The Surgeon General's Report on Health Promotion and Disease Prevention.* Washington, D.C.: U.S. Dept. of Health, Education and Welfare.

Van Harrison, R. 1978. "Person-Environment Fit and Job Stress." Chapter 7 in *Stress at Work,* edited by C. L. Cooper and R. Payne, 175–205. Toronto: John Wiley & Sons.

Vinokur, A., and M. L. Selzer. 1973. "Life Events, Stress, and Mental Distress." *Proceedings of the 81st Annual Convention of the American Psychological Association* (Montreal, Canada) 8:329–30.

Wagner, E. H. 1982. "The North Karelia Project: What It Tells Us about the Prevention of Cardiovascular Disease." *American Journal of Public Health* 72(1):51–53.

Wagner, E. H., et al. 1982. "An Assessment of Health Hazard/Health Risk Appraisal." *American Journal of Public Health* 72(4):347–52.

Wahlund, I., and G. Nerell. 1976. *Work Environment of White Collar Workers: Work, Health, Wellbeing.* Central Organization of Salaried Employees in Sweden.

Waldron, I. 1976. "Why Do Women Live Longer Than Men?" *Journal of Human Stress* 2(2):19–30.

Walker, K., and M. Shain. 1983. "Employee Assistance Programming: In Search of Effective Interventions for the Problem-drinking Employee." *British Journal of Addiction* 78:291–303.

Walters, V. 1982. "State, Capital and Labour: The Introduction of Federal-Provincial Insurance for Physician Care in Canada." *Canadian Review of Sociology and Anthropology* 19(2):157–72.

War, P., and T. Wall. 1975. *Work and Well-Being.* London: Penguin Books.

Ward, R. A., S. R. Sherman, and M. LaGory. 1984. "Subjective Network Assessments and Subjective Well-Being." *Journal of Gerontology* 39(1):93–101.

Weiner, H. J., S. H. Akabas, and J. J. Sommer. 1973. *Mental Health Care in the World of Work.* New York: Association Press.

White, T. H. 1979. "Human Resource Management—Changing Times in Alberta." A report prepared for Alberta Labour. (June).

Wilbur, C. S., and D. Garner. 1984. "Marketing Health to Employees. The Johnson and Johnson 'Live for Life' Program." In *Marketing Health Behaviour. Principles, Techniques and Applications,* edited by L. W. Frederiksen, L. J. Solomon, and K. A. Brehony. New York: Plenum Press.

Wiley, J. A., and T. C. Camacho. 1980. "Lifestyle and Future Health: Evidence from the Alameda County Study." *Preventive Medicine* 9:1–21.

Wilson, E. 1977. *Women and the Welfare State.* London: Tavistock.

Wilson, G. T., D. B. Abrams, and T. R. Lipscomb. 1980. "Effects of Intoxication Levels and Drinking Pattern on Social Anxiety in Men." *Journal of Studies on Alcohol* 41(3):250–64.

Windsor, D. B. 1980. *The Quaker Enterprise. Friends in Business.* London: Frederick Muller Ltd.

Wing, S. 1984. "The Role of Medicine in the Decline of Hypertension—Related Mortality." *The International Journal of Health Services* 14(4):649–66.

Wollert, R., and N. Barron. 1983. "Avenues of Collaboration." In *Rediscovering Self-Help: Its Role in Social Care*, edited by D. L. Pancoast, P. Parker, and C. Froland, 105–23. Beverly Hills, Calif.: Sage.

Woolfolk, R. L., et al. 1982. "Effects of Progressive Relaxation and Meditation on Cognitive and Somatic Manifestations of Daily Stress." *Behaviour Research & Therapy* 20:461–67.

Work in America Institute Studies in Productivity, Book 2. 1978. *Productivity and the Quality of Working Life*. Scarsdale, N.Y.: Work in America Institute.

Zimering, S., and J. F. Calhoun. 1976. "Is There an Alcoholic Personality?" *Journal of Drug Education* 6(2):97–103.

Zook, C. J., and F. D. Moore. 1980. "High Cost Users of Medical Care." *New England Journal of Medicine* 302:996–1002.

Index

About the Contributors

Eli Bay is the founder and director of The Relaxation Response Ltd., an organization that offers programs in practical stress management and maximizing performance. Mr. Bay is a consultant and seminar leader who since 1978 has been taking his unique synthesis of relaxation training skills into organizations in order to improve the health, performance, and work satisfaction of employees at all levels of the organization. He has gained an international reputation for the quality and effectiveness of his practical, hands-on approach, and for his grasp of training needs in a time of rapidly accelerating change.

Harry Hodgson has a Bachelors of Business Administration degree from Western Michigan University and holds a Masters of Education degree from Wayne State University. He is currently employed as Industrial Consultant with the Addiction Research Foundation of Ontario in Windsor, Ontario. He is a charter member of the Employee Assistance Society of North America (EASNA), a member of ALMACA (Association of Labor-Management Administrators and Consultants on Alcoholism) and of the Canadian Association of Suicidology.

L. Mark Poudrier, a graduate of the University of Toronto (O.I.S.E.), has been involved in designing, implementing, and evaluating employee assistance programs and voluntary assistance programs for more than ten years. A former center director with the Addiction Research Foundation, Dr. Poudrier is presently a private consultant to labor and management in both the private and public sectors.

Christine Shain has a Bachelors degree in anthropology from the University of Toronto. She has been active in numerous community development projects including the foundation of school-age daycare programs and parent advocacy groups. Currently she conducts research and provides briefs to various social action groups in Metropolitan Toronto.

James G. Simon is center director, Georgian Bay Center, Addiction Research Foundation. He was formerly employee assistance consultant, Peel Region; helped organize Peel Employee Assistance Council, and is presently secretary of that organization. From 1976 to 1978 Mr. Simon was seconded by ARF to Lifeline Foundation as manager. In 1975 he established and chaired the planning committee for INPUT 75 and then stayed on to chair INPUT 77. INPUT is a biannual Canadian National Conference on EAPs. Simon is currently the chairman of Employee Assistance Ontario and chairs the planning committee for Event '85 for that organization.

About the Authors

Martin Shain is head of the Occupational Research Program at the Addiction Research Foundation of Ontario. He holds an M.A. in jurisprudence from Oxford and a Diploma in criminology from Cambridge. He is author of *Employee Assistance Programs* with Judith Groeneveld (Lexington Books, 1980) and of numerous articles in the field. He serves on the editorial board of the newly constituted *Employee Assistance Quarterly,* is on the advisory board of the *EAP Digest,* and belongs to the Employee Assistance Society of North America (EASNA).

Helen Suurvali is senior research assistant in the Occupational Research Program at the Addiction Research Foundation of Ontario, where she has been working since 1974. She has an honors Bachelors degree in Psychology from Carleton University (Ottawa). Since joining the A.R.F., Helen has been involved in evaluations of numerous health-related interventions including Parent Communication Training, alcoholism treatment and more recently EAPs and health promotion programs.

Marie Boutilier holds an honors B.A. in sociology from the University of Western Ontario, and an M.A. in sociology from McMaster University. Her areas of specialization include industrial sociology, and women's studies, with a special interest in the sociology of popular culture.